HAPTICS

D1043864

The MIT Press Essential Knowledge Series

HAPTICS

LYNETTE A. JONES

The MIT Press | Cambridge, Massachusetts | London, England

This book was set in Chaparral Pro by Toppan Best-set Premedia Limited. Printed and bound in the United States of America.

Library of Congress Cataloging-in-Publication Data

Names: Jones, Lynette A., author.
Title: Haptics / Lynette A. Jones.
Description: Cambridge, MA : The MIT Press, 2018. | Series: The MIT Press essential knowledge series | Includes bibliographical references and index.
Identifiers: LCCN 2018005636 | ISBN 9780262535809 (pbk. : alk. paper)
Subjects: LCSH: Touch. | Touch--Physiological aspects. | Senses and sensation.
Classification: LCC QP451 .J66 2018 | DDC 612.8/8--dc23 LC record available at https://lccn.loc.gov/2018005636

10 9 8 7 6 5 4 3 2 1

CONTENTS

SERIES FOREWORD

The MIT Press Essential Knowledge series offers accessible, concise, beautifully produced pocket-size books on topics of current interest. Written by leading thinkers, the books in this series deliver expert overviews of subjects that range from the cultural and the historical to the scientific and the technical.

In today's era of instant information gratification, we have ready access to opinions, rationalizations, and superficial descriptions. Much harder to come by is the foundational knowledge that informs a principled understanding of the world. Essential Knowledge books fill that need. Synthesizing specialized subject matter for nonspecialists and engaging critical topics through fundamentals, each of these compact volumes offers readers a point of access to complex ideas.

Bruce Tidor
Professor of Biological Engineering and Computer Science
Massachusetts Institute of Technology

HOW WE PERCEIVE THE WORLD VIA TOUCH

When we move our fingertips across a table or along a piece of fabric we are immediately able to sense whether it is smooth or rough, and even in the absence of vision we can probably determine what the table or cloth is made from. We can also recognize familiar objects quickly and very accurately using touch alone. This ability to identify and perceive the properties of objects relies on our sense of touch and more specifically on active touch, which is often referred to as haptics or haptic sensing. The word haptics has its origin in the Greek word *haptikós*, meaning "able to grasp or perceive."

Haptic sensing is critical to our experience of the world, from providing us with information that enables us to use just the right amount of force to lift a glass of water from a table, to finding the light switch on the bedroom wall in the dark. In both these situations, there

Haptic sensing is critical to our experience of the world, from providing us with information that enables us to use just the right amount of force to lift a glass of water from a table, to finding the light switch on the bedroom wall in the dark.

is ongoing feedback from sensory receptors in the skin and muscles so that we can immediately make an adjustment if we notice that the glass is slipping between our fingers, or that our hand has not located the light switch after scanning the surface of the wall. In the absence of this feedback, we are clumsy and rely excessively on vision to guide our hand movements. We are usually unaware of how essential active touch or haptics is to our daily experience until something limits or interferes with the way we perform a task. When our hands are numb due to cold or very sweaty in hot weather, we are forced to be much more cautious when we handle objects or perform an action, because the sensory signals coming from the hand are different from those we normally experience.

Although everyone is familiar with the sense of touch, the word *haptics* is less frequently used and so is unfamiliar to many people. It involves two senses, touch and kinesthesia, the latter referring to the sense of limb position and movement. Because we primarily use our hands to explore the world tactually, haptic sensing is intimately related to the function of the hands and how we use them. Unlike the visual system, where information such as the color and shape of an object is available with minimal effort, to acquire haptic information we must move our fingers across an object's surface to perceive its texture or poke at it with our fingers to determine if it is hard or soft. Much of the early scientific research on the sense of touch did not focus on active haptic exploration, but simply presented a force or vibration on the skin of a passive observer, who was required to report what was felt.

The aim of this book is to provide an overview of the many aspects of active touch sensing, from the sensors in our skin and muscles that enable us to perceive the world, to the specialization of the haptic sense for processing information about the material properties of objects. Such information is essential to developing robotic and prosthetic hands that attempt to mimic the properties of human hands. We also consider how the sense of touch can be used to compensate for the loss of function in other sensory modalities, like vision and audition. The book will also explore the nature of new technologies that are being developed to create tactile surfaces on the flat screen displays that are so pervasive in consumer electronic devices.

The focus of the first chapter is on understanding the building blocks of the tactile system, namely, the sensory receptors in skin that signal contact with an object and the overall structure of the skin itself. By understanding the properties of these receptors and the skin, we can appreciate how information is processed at this interface with the external world.

Introduction

Haptics involves integrating information from two senses: touch and kinesthesia. The tactile cues arise from stimulation of sensors in the skin known as mechanoreceptors. Kinesthetic information comes from the sensors in muscles, tendons, and joints that signal the position and movement of our limbs, and the forces generated by muscles. The term *kinesthesia* is often used interchangeably with *proprioception*, although the latter is usually considered to include information from the vestibular system, which is involved in controlling posture and balance, along with the sensory information from muscles, tendons, and joints. The essential element of haptics and haptic exploration is that there is active movement of the hand, so that the sensory information a person receives does not come just from passive contact but from actively exploring the environment. This distinction is important because the knowledge we get when we squeeze a peach to see if it is ripe, or lift a package to determine its weight, is usually much more detailed and precise than what we can obtain if the peach or package is placed on our outstretched hand resting on a table. Haptic sensing therefore differs from other senses, like vision and audition, in that it is bidirectional: the information we can extract about an object's properties is exquisitely linked to the movements made to perceive those properties. In the act of exploring an object

Haptic sensing therefore differs from other senses like vision and audition, in that it is bidirectional; the information we can extract about an object's properties is exquisitely linked to the movements made to perceive those properties.

we may even change its properties, such as when we exert too much force on a ripe strawberry and crush it, permanently altering its shape. Such irreversible interactions do not occur in vision or audition—the act of looking at a tree or listening to music does not change its state. As a sensory system, haptics is remarkably efficient at processing the material characteristics of objects, for example, their surface textures, and is the sensory system of choice when making such judgments. If we want to determine whether a scratch on a table has marred the surface or is just a superficial mark, we will usually move our fingers along the table to feel the depth of the scratch rather than simply relying on visual cues.

Over the past decade, the field of haptics has emerged beyond the purview of engineers and scientists to a much broader audience interested in re-creating the tactile and haptic feedback now absent in their interactions with smartphones, tablets, and virtual and augmented reality systems. The challenge here is to reproduce in a virtual world some of the sensations associated with normal physical interactions with the environment, like the warmth and softness of human skin or the vibrations associated with driving over a potholed road. With the advent of more sophisticated and higher resolution head-mounted displays, such as HoloLens (Microsoft), Gear VR (Samsung), and Oculus Rift (Facebook), comes the need to add a physical dimension to the visual interactions, particularly when these systems are used for education and training. In addition, since so many of our interactions are with touch screens, there is growing recognition of the importance of creating virtual haptic effects on these physical surfaces—an area known as *surface haptics*. Finally, with the growth of wearable technologies there is renewed interest in using the skin as a medium of communication, to offload the overworked visual and auditory sensory systems. The focus of such work is to determine whether we can develop tactile displays that have greater potential for communication than the simple vibration signals currently implemented as alerts and notifications in many devices. More sophisticated communication systems based on vibrotactile patterns have yet to be developed, although the functionality does exist in some devices to create them.

Tactile System

The sensory modality that relates to the body senses is often referred to as the somatosensory system, although it is really made up of a number of submodalities. A broad distinction can be made between sensory information that originates in the skin and that which comes from muscles and joints. However, even within the skin itself there are four recognized submodalities, often collectively referred to as the cutaneous senses, illustrated in figure 1. The skin senses are touch, temperature, pain, and itch. It has been proposed that a fifth modality exists that conveys the positive, affective (pleasant) properties of touch.[1]

Touch and temperature contribute to the discriminative functions of the skin by providing us with information about temporal and spatial events on the body. Based on the sensory signals that come from tactile and temperature receptors in the skin, we can find a coin in our pocket or know whether the cup of coffee in our hands may be too hot to drink. In contrast, pain receptors protect the skin against potential or actual damage and can respond to thermal, mechanical, or chemical stimuli. The pain sensory system is referred to as nociception, and within this submodality of the somatosensory system some receptors are specialized for detecting particular kinds of tissue-damaging inputs, such as the high temperature of a hot plate or the sharp edge of a knife. Another class of pain receptors are called polymodal receptors; they respond

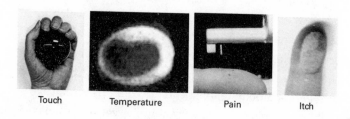

Touch   Temperature   Pain   Itch

Figure 1 Submodalities of the skin sense.

not only to intense mechanical stimulation, for example a pinprick, but also heat and noxious chemicals, like the capsaicin found in chili peppers. The final cutaneous sense is itch, which was thought in the past to be part of the pain-sensing system but is now considered to be a distinct submodality, in that the pathway from the skin to the brain that mediates itch is independent of those involved in nociception. Itch can be a transient sensation triggered by mechanical stimuli like the coarse wool of a sweater or by chemical stimuli. The latter includes the inflammatory agents released by plants like poison ivy and poison oak when they come in contact with the skin, and the clear liquid in the vesicles (red spots) that form on the skin with chickenpox. Itching can also result from damage to sensory nerves or the brain.

Sense of Touch

The human sense of touch is housed in the largest and heaviest of all the sense organs—the skin—which in an average adult covers an area of about 1.8 m^2 and weighs around 4 kg (about 9 lb). Apart from its sensory function, the skin acts as a protective barrier and is also involved in regulating body temperature. It contains more than 2 million sweat glands, which produce sweat to prevent the body from overheating, and around 5 million hairs. There are two main types of skin: *hairy skin*, which covers most of the body including the back or dorsal surface of the hand, and hairless or *glabrous skin*, which covers the palmar surface of the hands and the plantar surface of the feet. The skin on the lips and genitalia is also hairless but differs from the glabrous skin on the palmar and plantar surfaces: the skin on the lips does not have sweat glands, and on the genitalia there is a mucosal surface.

Hairy skin contains both soft, fine vellus hairs that serve a wicking function in drawing perspiration away from the skin and assist in thermal insulation, and the longer, thicker, and more visible guard hairs. Both types of skin have the same overall structure, with an outermost layer of dead skin cells called the *stratum corneum*, and then three sublayers that contain several types of cells including keratinocytes, Langerhans cells, and melanocytes. Collectively these four layers of skin are known as the epidermis. Keratinocytes comprise about 90% of the cells in the epidermis and serve as a barrier against environmental damage; Langerhans cells protect against infection and so are part of the immune system, whereas melanocytes manufacture the pigment melanin, which determines the color of skin. Beneath the epidermis is the dermis, which contains sweat glands, blood vessels, sensory nerve endings, and a network of collagen fibers.[2]

On the palm, glabrous skin is relatively thick but can bend along the flexure lines of the hand when an object is grasped. These lines are visible in your palm when you flex all your fingers together. The skin here is tethered by fibrous tissue that connects the deep layers of the skin with the sheaths of the underlying tendons that flex the fingers, and so they are regions that remain relatively fixed during movements of the hand. Papillary ridges are found all over the skin on the palm in various configurations, including the curved parallel patterns on the fingertips that provide the basis of the loops, whorls, and arches of fingerprints (see figure 2). The patterns and characteristics of single ridges in fingerprints are permanently fixed and, except for enlargement as a function of maturation, there is no change in the ridge characteristics of a finger pattern throughout the lifetime of an individual, hence their use in identification and forensic science. Fingerprints are found in other mammalian species including chimpanzees and gorillas and, surprisingly, koalas. It is thought that

Anti-slip mechanism
Assist in transmitting pressure cues
House sweat glands, making skin adhesive

Figure 2 Functions of the papillary ridges on the fingertips.

their functional significance across species may relate to grasping.

Papillary ridges serve several functions that are important to the sense of touch and grasping. First, similar to the tread on a tire or the milled grooves on the handle of a tool, they serve as an anti-slip mechanism and so contribute to grasp stability. Second, they assist in transmitting information about pressure or contact on the skin surface to the nerve endings of sensory receptors that lie directly beneath them. Third, the ridges house eccrine sweat glands that are just below the epidermis, the outermost layer of skin. Sweat glands keep the skin moist and adhesive, which greatly facilitates grasping. When the skin sweats, the papillary ridges swell and become slightly elevated, making them more effective points of contact when holding an object.[3] In contrast to the glabrous skin on the palm of the hand, the skin on the dorsal surface of the hand is thin, soft, and pliable because it lacks the connective tissue that anchors the palmar skin. When the fingers move, the skin on the back of the hand stretches and can be lengthened by up to 30 mm (over an inch) as a finger moves from extension to full flexion.

Tactile Receptors

At various depths under the skin surface are different types of sensory receptors that provide information to the central nervous system and the somatosensory areas of the brain about mechanical events occurring on the skin surface. Based on signals coming from these receptors we are aware when a fly has landed on our skin or when a glass held in the hand is beginning to slip from our grasp. Tactile receptors are found all over the body, in both hairy and glabrous skin, and are embedded in both the epidermis and the underlying dermal layer.[4] There are differences between these two types of skin in terms of the receptors they house and the inputs that trigger a response. In the glabrous skin there are four types of tactile receptors, named after the anatomists who first described them: *Meissner's corpuscles*, *Pacinian corpuscles*, *Merkel cells*, and *Paciniform endings*. As their names suggest, they are structurally distinct and are distinguished based on (1) the size of the area of skin that when stimulated produces activity in the sensory afferent fiber connected to the receptor, and (2) their responses to sustained indentation of the skin. The responsive area of skin is known as the *receptive field* and is categorized as being either type I (small: 2–8 mm in diameter) or type II (large: 10–100 mm diameter). The other characteristic used to classify mechanoreceptors is the dynamics of their responses, or more precisely their *rate of adaptation*. Fast-adapting (FA) receptors signal when the skin is actively being indented but do not respond when movement of the skin stops, even if it remains indented. These receptors are important in sensing very small movements of the skin, such as the initial contact of an object with the skin when grasping. In contrast, slowly adapting (SA) mechanoreceptors respond both while the skin is moving and during the period of sustained indentation. They are involved in perceiving shape, coarse textures, and the direction of motion of an object along the skin.

The distribution of mechanoreceptors varies over the body, with those having a small receptive field being concentrated in the fingertips and becoming progressively less numerous as we move proximally toward the wrist and forearm. Mechanoreceptors with large receptive fields are more sparsely distributed and do not show such a marked variation in density on the hand. Different types of mechanoreceptors respond to different ranges of frequency of vibration applied to the skin. Collectively, mechanoreceptors respond to vibration on the skin from about 0.4 Hz to 1000 Hz. The SA type I units are maximally sensitive to very low frequencies of vibration, between 0.4 and 3 Hz, whereas the FA I units respond to frequencies between 1.5 and 100 Hz and the FA II units respond to higher frequencies, between 35 and 1000 Hz.[5]

The distribution of mechanoreceptors over the skin surface and the sizes of their receptive fields determines the spatial acuity of the skin. Areas that have a high density of tactile receptors, such as the fingertips, are much more sensitive to mechanical stimuli than areas with lower densities, like the palm. The greater sensitivity of the fingertips is reflected in the very small displacements (10 μm) and forces (10 mN) that are perceptible when applied to the skin. This variation in tactile acuity underlies another fundamental difference between touch and vision and audition, namely that the sense of touch is distributed across the body. This means that the ability to perceive a particular tactile stimulus will vary across the skin surface, and that the same tactile stimulus will not be perceived identically at different locations.

When we manipulate an object in our hands, all types of mechanoreceptors respond as the skin is mechanically stimulated. However, different types of receptor are critically important for certain tactile functions. For example, the ability to perceive fine spatial features, such as the edges of a plate or the curvature of a baseball, depends on slowly adapting type I (SA I) receptors that have high sensitivity to the temporal and spatial aspects of stimuli. In contrast, fast-adapting type I (FA I) receptors respond when there is movement between the skin and an object; this is important for detecting slip and maintaining a steady grasp.

In hairy skin, there are five main types of mechanoreceptors, two of which are slowly adapting and three are fast adapting. The slowly adapting receptors are Merkel cells and Ruffini endings, and the fast-adapting receptors include hair-follicle receptors that innervate hairs, field units, and Pacinian corpuscles. The receptors found around hair follicles respond to bending movements of individual guard and vellus hairs. One type of receptor can detect very small deflections of hairs and responds in a different manner to hair deflection toward the skin surface as compared to away from it. Another type of receptor is sensitive to axial movement of hairs, such as occurs when they are pulled. There are also receptors in hairy skin, not found in glabrous skin, that have low thresholds to mechanical stimuli and whose signals are transmitted via slow-conducting (unmyelinated) nerve fibers to the brain. These receptors and associated afferent units are known as *C-tactile (CT) afferents* and respond to slow, gentle movements across the skin, such as slow stroking of the skin during intimate interactions. The firing rate of these afferent units is positively correlated with people's ratings of touch pleasantness. The properties of these receptors, such as their slow conduction velocities, indicate that they are unlikely to be involved in rapid tactile discrimination or cognitive tasks. An interesting feature of this class of mechanoreceptor is that they appear to be tuned to the temperature of the stimulus stroking the skin. They respond more vigorously to slowly moving stimuli at a neutral (skin) temperature than at cooler or warmer temperatures. The information from these receptors is thought to be involved in the affective or pleasant aspects of touch, since their responses appear to be optimized for signaling pleasant skin-to-skin contact in humans.[6]

Our understanding of how these various mechano-receptors respond to mechanical stimulation of the skin comes in part from neurophysiological studies of normal, healthy people. The neural activity from isolated sensory fibers in the median nerve in the forearm can be recorded using fine tungsten microelectrodes. The electrodes are manually inserted into the skin, penetrating the under-lying tissue until a nerve fascicle is found and impaled. The nerve fibers are typically connected to a number of mechanoreceptors in the skin of the hand, and they dis-charge (send a signal to the central nervous system) when the mechanoreceptor is stimulated. The presence and fre-quency of discharges recorded from the nerve fiber indi-cate whether the receptor responds to a specific form of stimulation—for example, constant pressure—or to the frequency of vibration of a probe on the skin. Based on whether these mechanoreceptors respond to such stimuli and on the properties of the responses, various neural coding schemes have been proposed to account for human tactile perception.[7] Such research seeks to understand the relation between the responses of a number of mechano-receptors and their associated neural units, and a percep-tual entity such as the perceived roughness of a surface or shape of an object (see figure 3).

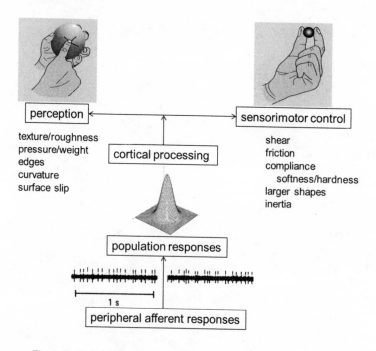

Figure 3 Pathway from mechanoreceptors in the skin to the brain, where the afferent signals are used for both perception and sensorimotor control of the hand. From Jones and Smith (2014)[7] with permission from Wiley Periodicals, Inc.

SENSORY AND MOTOR
SPECIALIZATION OF THE HAND

This chapter provides an overview of the basic properties of kinesthesia. Those properties, in conjunction with the tactile sensing capabilities described in chapter 1, delineate the features of the haptic sense. We will also consider the remarkable haptic sensing abilities of nonprimate species, such as rodents and moles, which use whiskers or fleshy appendages around their snouts to explore their environments tactually. The speed with which these animals are able to identify surfaces and prey, based on haptic cues, is truly extraordinary. In addition to receptors that respond to pressure and movement on the skin surface, there are sensors in the skin that detect changes in temperature. The properties of such cutaneous thermoreceptors are reviewed, together with their role in helping us to identify the material composition of objects in contact with the skin. Finally, the structure of the hand and associated musculature is described, along with the role of tactile signals in controlling hand movements.

Kinesthesia

Kinesthetic information comes from sensory receptors in muscles, called muscle spindles, that provide the central nervous system with information about the length of muscles and the rate at which muscle length changes. These sensory signals enable us to perceive the direction, amplitude, and velocity of limb movements and the associated changes in limb position. Cutaneous mechanoreceptors also contribute to the perception of limb movement, via their responses to stretch of the skin overlying the moving joint. When a joint moves—for example, when we extend our index finger to point at something—the skin on the palmar side of the joint is stretched and on the back of the hand it becomes slackened or folded. The stretch of the skin is sensed by slowly adapting cutaneous mechanoreceptors that signal to the central nervous system the position of the joint. The sensitivity of skin stretch receptors has been shown to be similar to that of muscle spindle receptors, when expressed in terms of the number of impulses recorded in a sensory nerve per degree of joint motion.

Receptors in and around joints play a more minor role in kinesthesia and appear to act primarily as limit detectors, signaling when a joint is near an extreme position such as maximum extension or flexion. Much of the time we are unaware of kinesthetic signals, even though we can indicate with some degree of accuracy with our eyes closed the position of our limbs and whether they are moving or stationary. Our attention is drawn to these signals when a movement does not go as planned, such as when we encounter an obstacle as we extend our arm across a desk to pick up a pen. In these situations, the incoming sensory feedback differs from what was anticipated based on the outgoing, efferent motor command sent from the brain to the muscles, and so corrective action must be taken.[1]

The critical importance of kinesthetic receptors to normal function has been highlighted by the case of an individual (IW) who at the age of 19 was suddenly affected by a rare neurological disease that eliminated all tactile and kinesthetic feedback from his peripheral nerves, depriving him of any feeling from his neck to his feet. Over a period of nearly two years in hospital in which there was no neurological recovery, IW taught himself to walk, dress, and feed himself by thinking about the movement, and then visually monitoring the moving limb. Even after 30 years of executing successful movements, he still has to concentrate and monitor every movement visually to perceive where his limbs are, since he continues to lack any feedback from his peripheral nerves. It is impossible for him to stand or control the movements of his arms and legs without this deliberate and conscious process in which the consequences of every movement have to be assessed. If the lights accidentally go out at night he collapses to the floor, not realizing what has occurred until the lights come on again. This individual's remarkable ability and resilience in learning how to compensate for his profound loss is documented in two books written by a clinical neurophysiologist who has worked with him. *Pride and a Daily Marathon*[2] covers the early years of IW's loss and attempts to overcome his impairments, and the more recent account of his life since this early period is in "*Losing Touch: A Man Without His Body.*"[3] He is also the subject of a BBC documentary film, *The Man Who Lost His Body*. The extent of IW's functional recovery is highly unusual for someone with such a debilitating condition; his level of function is at the cost of a complete reliance on visual cues when moving his limbs, and the associated tremendous cognitive load, as reflected in the title of the book, describing life as "a daily marathon."

In addition to providing information about the position and movements of our limbs, the kinesthetic sense is involved in the perception of force. The forces generated by muscles during voluntary movements are sensed by receptors, known as Golgi tendon organs, that are found at the junction between muscle fibers and collagen strands of tendons. An additional source of force information comes from the motor command sent from the brain to produce contractions in muscles. Correlates, or copies, of these commands are transmitted to the sensory areas of the brain to facilitate the processing of incoming feedback from the muscles. Information about the forces generated by muscles can be sensed from these motor command correlates, known as *corollary discharges* or efference copies, which reflect the magnitude of the neural signal sent to a muscle. Sensing force, therefore, depends on contributions from feed-forward pathways in the brain and feedback pathways from the periphery.[4] The sources of information for each perceptual component of kinesthesia are listed in table 2.1.

For the tactile system, higher densities of mechanoreceptors, as found in the fingertips and around the mouth,

Table 2.1 Sensory basis of kinesthetic perception

Sensory event	Source of information
Perception of limb movement	Muscle spindle receptors Cutaneous mechanoreceptors Joint receptors
Perception of limb position	Muscle spindle receptors Cutaneous mechanoreceptors Joint receptors
Perception of force	Golgi tendon organs Corollary discharges

are associated with superior tactile acuity. This means that the threshold for detecting the height of a small dot moving across the skin, or the distance between two points in contact with the skin is much smaller in areas with higher densities of receptors. For the kinesthetic system this does not appear to be the case. The overall number of muscle receptors is much smaller, around 25,000 to 30,000 muscle spindles in the human body, with 4,000 in the muscles in each arm, as compared to 17,000 cutaneous mechanoreceptors in the human hand alone. Higher densities of muscle spindle receptors are not clearly associated with superior kinesthetic abilities. Muscle spindle density appears to depend more on the size of the muscle than its function. The number of spindles in human muscles varies, and for muscles in the arm it has been estimated to range from 34 for the first dorsal interosseous, a muscle found in the back of the hand that is involved in abducting the index finger (moving it away from the midline of the hand), to 320 in the biceps brachii, a muscle that flexes the elbow.[5] When expressed in terms of the number of spindles per gram of mean weight of adult muscle, high spindle densities are found in the muscles within the hand itself, known as the intrinsic hand muscles, and the highest densities are found in the deep layers of neck muscles where densities of up to 500 spindles per gram have been reported.

When we look at the structure of the hand and the wealth of sensory information that arises even during the most mundane of activities, like picking up a pen, the challenge of deciphering what a particular neural input means seems formidable. The muscles that move the fingers lie in both the forearm and the hand, and for many muscles their tendons cross three or more joints. This means that the signals arising from muscle spindle receptors are often ambiguous rather than providing specific information about which joints have moved. A similar ambiguity can occur for signals from cutaneous receptors, in that many receptors respond to movements of more than one finger. However, from the spatial array of inputs that arise from a number of different types of receptor, the central nervous system can compute the position and movements of various joints. This convergence of afferent information and its processing occurs both within the spinal cord and at higher levels in the brain.

Haptic Sensing in Other Species

Much of what we know about cutaneous mechanoreceptors and the underlying mechanisms involved in haptic processing comes from research on the hands of human and nonhuman primates, particularly the fingertips, which are one of the most tactually sensitive locations on

When we look at the structure of the hand and the wealth of sensory information that arises even during the most mundane of activities, like picking up a pen, the challenge of deciphering what a particular neural input means seems formidable.

the body. For other species, such as rodents, the focus of tactile research is on the vibrissae or whisker system. Vibrissae differ from ordinary hair in that they are longer and thicker and have large follicles at their base that contain blood-filled sinus tissues.[6] A rodent uses a set of approximately 30 whiskers on each side of its snout to palpate surfaces, using forward and backward motions known as whisking (see figure 4). Movements of the whiskers are coordinated with those of the head and body, allowing the animal to locate stimuli of interest through whisker contact. Items of interest can then be explored further using the vibrissae and shorter, nonactuated microvibrissae on the chin and lips. These whisking movements are fast and can occur at rates of up to 5 to 15 times per second. When the tip or shaft of the whisker makes contact with a surface, the movement changes. This change in motion is detected by sensory receptors that signal to the somatosensory areas of the brain where the contact was made. Based on these sensory signals rodents can identify a surface texture rapidly (within 100 ms of contact) and with remarkable accuracy.

Another animal that shows extraordinary tactile abilities and uses the sense of touch to navigate its environment is the star-nosed mole (figure 4). It spends its life in subterranean environments, in complete darkness, and is functionally blind. On either side of its snout are 11 pairs of fleshy appendages on which there are 25,000

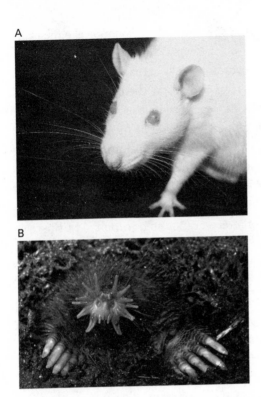

Figure 4 (A) Whiskers used by the rat to explore its environment (Courtesy Mitra Hartmann, Northwestern University). (B) Star-nosed mole with its fleshy appendages used to locate and identify prey (from Gerhold et al., 2013).[7]

receptors known as Eimer's organs, a class of tactile receptor found in nearly all other species of moles, but at substantially smaller numbers. The star configuration of the nose is unique in the animal kingdom; it is in constant motion as the mole explores its environment to identify food. When foraging the mole can touch the star to the ground 10 to 15 times per second. It is possible for the mole to identify whether prey is edible or not within 8 ms, which is basically the limit of the speed of its neuronal processes. The star-nosed mole is known as the "fastest-eating mammal"—it can identify and consume small prey in 120 ms.[8]

An understanding of how humans and other animal species sense the environment haptically is important to the design of artificial sensing systems such as those used in robotic and prosthetic hands (see chapter 8). Replication of human sensing capabilities presents many challenges, particularly given the large number of receptors found in the hand and the robustness of the skin. The human hand can be immersed in liquids, withstand heat and cold, generate considerable forces when grasping heavy objects, and undergo considerable strain when the skin is stretched, all without damage to the skin or underlying receptors. Designing and fabricating artificial sensors that are as robust and versatile as those found in the hand is a daunting engineering task. Robotic hands require sensors that perform similarly to those found in the skin, and

It is possible for the mole to identify whether prey is edible or not within 8 ms, which is basically the limit of the speed of its neuronal processes. The star-nosed mole is known as the "fastest-eating mammal"—it can identify and consume small prey in 120 ms.

motors that can respond rapidly to changing sensory input based on the conditions of manipulation.

Thermal Sensing

When we hold a tool, we are often able to identify whether it is made from metal or plastic based on changes in skin temperature. These thermal cues become important when we have to identify objects in the dark or when our visual attention is directed elsewhere, for example while driving a car. Changes in skin temperature are sensed by thermoreceptors located in the epidermal and dermal layers that respond to either increases (warm thermoreceptors) or decreases (cold thermoreceptors) in temperature. These receptors are essential for maintaining our core body temperature within a narrow range of ±0.5°C, even while skin temperature may vary from around 28°C to 36°C (82°F–97°F) under normal conditions. The two types of temperature receptors are independently distributed over the body, and cold receptors outnumber warm receptors. Even though the distribution of cold and warm thermoreceptors varies across the skin surface, all body regions are more sensitive to cold than to warmth. When the temperature of the skin increases above 43°C or decreases below 15°C, pain receptors in the skin signal these tissue-damaging temperatures to the central nervous system.[9]

In contrast to the sense of touch, the fingertips are not the most thermally sensitive region of the hand; the area at the base of the thumb, known as the thenar eminence, and the skin on the back of the hand are more sensitive to changes in temperature. Although we may not be consciously aware of these differences in sensitivity, it is interesting to note that if we want to check the temperature of a child who may have a fever or of the formula in a bottle that is being given to a baby, we will often use the back of the hand or the wrist for this purpose.

When certain chemical compounds are placed on the skin they are perceived as being hot or cool, even though they are at room temperature. Mint—and in particular menthol, which is the main active ingredient in mint—is perceived to cool the skin, whereas chili peppers, whose active ingredient is capsaicin, feel hot when placed on the skin. These perceptual effects result from sensory processes in the skin, specifically involving the free nerve endings in the epidermal layer. It has been found that there are specialized sensors in the nerve endings that respond to both heat and capsaicin, and a different class of sensory nerve ending that sends a signal to the central nervous system when the skin is cooled or menthol is in contact with the skin surface.[10] These substances and related compounds are used in many folk and nonprescription remedies for their analgesic effects. Capsaicin is found in topical creams to stimulate blood flow and relieve mild muscle pain, whereas menthol is used to produce a sensation of pleasant cooling on the skin surface.

The changes in skin temperature that occur when we grasp an object and help us identify what it is made from depend on the thermal properties of the object and on the initial temperatures of the skin and the object. Generally, the resting temperature of the skin is higher than the temperature of objects encountered in the environment, so it is the decrease in skin temperature on contact—or heat transfer from the skin to the object—that helps us identify it. The thermal properties of materials like copper and stainless steel, with their high thermal conductivity and low specific heat capacity, mean that there is a higher rate of heat flow from the skin to the metal object, as compared to contact with an object made from plastic or rubber.

Thermal cues not only assist us in identifying objects but can also have an effect on haptic perception. When people are asked to judge the heaviness of two items of equal weight but different temperatures, the colder item is perceived to be heavier than the item maintained at skin temperature. These effects can be substantial, with a very cold, light weight of 10 g being perceived to be perceptually equivalent to a weight of 100 g maintained at skin temperature. Warming an object also makes it feel heavier than an object at skin temperature, but the effect here is much smaller and varies across people. These temperature-weight interactions are presumed to result from temperature-related changes in the sensitivity of mechanoreceptors in the skin that signal pressure.

Thermoreceptors are also involved in the perception of wetness and humidity, for which humans, unlike insects, do not have specialized humidity sensors known as hygroreceptors. We are aware when a drop of water falls on our skin, or when our hand is immersed in a liquid and can perceive varying levels of moisture in wet fabrics. Thermal and tactile cues are integrated to give rise to these sensations. When the hand is immersed in a liquid, mechanical pressure cues on the skin enable us to perceive the ring of pressure at the interface between the liquid and air. Coupled with these sensations is an awareness of the heat that is usually conducted away faster from the immersed skin as compared to the skin in air, a signal that arises from thermoreceptors.[11] The ability to perceive varying levels of moisture or wetness in textiles appears to be partly based on mechanical cues such as stickiness. This is reflected in the improved ability to discriminate different degrees of wetness in textiles if the fingers are able move across the fabric, as compared to when they rest passively on the surface.

Motor System

Unlike other sensory systems, the haptic system is bidirectional, in that the information it receives is a function of the hand movements used to explore the environment. If those movements are constrained in some way, by limiting the motion of the fingers with splints, wearing gloves, or only allowing exploration with a tool such as a stylus rather than a bare finger, the information extracted from the environment will be inferior to that obtained under normal exploratory conditions. It is of interest therefore to consider how the structure of the human hand influences manual exploration.

The five-fingered hand has been referred to as the "absolute bedrock of mammalian primitiveness," a statement that reflects the finding that no other mammal, reptile, amphibian or bird has ever evolved a form with more than five digits.[12] When we look at a skeleton of the human hand we can see that it is composed of 27 bones, 14 of which are phalanges, or finger bones, 5 are metacarpals, or palm bones, and 8 are carpals, or wrist bones. They all articulate with at least one other bone via a synovial joint, the most common type of joint in the body. The finger joints are hinge joints, which allow one axis of movement, flexion or extension, whereas the joints at the base of the fingers where the proximal phalanges and metacarpals articulate are biaxial or condyloid joints with two axes of motion: flexion and extension, and movement away from and toward the midline of the hand (abduction/adduction). In total, the human hand including the wrist has 21 degrees of freedom of movement. Movements of the fingers are not, however, completely independent, as you can see when you try to move only your little finger and find that the ring finger also moves slightly. There are soft tissues between adjacent digits that result in passive biomechanical coupling of the fingers, as well as interconnections between the tendons of some muscles that attach to multiple fingers, both of which contribute to the coupling of movements.

One of the most important joints of the hand from a functional perspective is the saddle joint at the base of the thumb. This joint permits flexion/extension, adduction/abduction, and medial-lateral rotation, allowing the thumb to rotate and make contact with the tips of the fingers in what is known as a *precision grip*. The ability to oppose the thumb against the fingers is essential for handling and exploring small objects. From an evolutionary viewpoint, the opposable thumb greatly facilitated primate tool use and tool construction. Although many Old-World monkeys and primates have opposable thumbs, the extensive area of contact between the tip of the index finger and the thumb is a unique human characteristic. In the absence of a thumb, following injury or amputation, it is estimated that the hand has lost 40% of its function. This loss is of such significance to daily living that reconstructive surgery involving transferring the great toe from the foot to the hand is occasionally performed in order to give the hand some opposable capacity.[13] The functional independence of the thumb makes it the most specialized of all the digits, and although most primates have some degree of functional independence of this digit, humans are the most skilled in its use.

Most of the muscles that produce movements of the fingers lie in the forearm and have long tendons that extend from the region of the elbow to the fingers. These are known as the extrinsic hand muscles. Smaller muscles within the hand itself (intrinsic) are also involved in some movements. One class of these muscles, known as the lumbricals, is unique in the human body in that the muscles originate from tendons in the hand and not from bone. There are 29 muscles that control hand movements, although some of these muscles are divided into distinct parts with separate tendons. If the subdivisions of the muscles are counted separately, the number of muscles controlling movements of the hand increases to 38, an impressive number when one considers the task of replicating the hand's functionality in robotic or prosthetic devices.

Sensorimotor Control of the Hand

Signals from mechanoreceptors in the skin are not only involved in tactile perception but are essential for the control and modulation of force when we manipulate objects (see figure 3). This distinction is often referred to as *action for perception*—perceiving the world outside our bodies—and *perception for action*—controlling our manual interactions with the world. Sensory signals from mechanoreceptors play a critical role in encoding the magnitude, direction, timing, and spatial distribution of fingertip forces. This is important because the forces we use to explore a surface during active touch, or to grasp a fragile object without damaging it, are optimized to perform the task. Grip forces are adjusted as a function of the weight of an object and the friction between the skin and the object. It is important that the forces used to grasp the object are such that it is stable in the hand and does not easily slip between the fingers. The term *safety margin* is used to describe the difference between the force at which the object begins to slip between the fingers and the force used for grasping. If the safety margin is large the muscles controlling the fingers can fatigue, and if it is too small the object may slip. If an object does begin to slip between the fingers there is an automatic increase in grip force within 70 ms, which results in a more stable grasp. These slips are not usually perceived even though they are detected by mechanoreceptors in the skin on the finger pad that signal to the central nervous system the need to increase grip force.[14]

When tactile feedback is deficient or absent we cannot control grip forces adequately and so we are unable to compensate when an object begins to slip between our fingers or to adjust the forces when friction changes. The forces used to grasp an object when the fingers are anesthetized are usually much larger than normal and are no longer optimized for the weight of the object and the friction between the skin and object. In everyday life, we are aware of the importance of tactile feedback in the control of our movements when our fingers are very cold and become numb, or when we have to wear thick gloves on our hands. In these situations, we have difficulty in performing the most basic of functions, such as inserting a key into a lock or picking up a paper clip from a desk.

HAPTIC PERCEPTION

The sensory information from receptors in the skin and muscles is transmitted to the brain, where it is processed in various cortical areas to give rise to our haptic experiences. This chapter examines the haptic sense with a focus on how it is specialized to perceive the material properties of objects, and so is the sense of choice when judgments about such features have to be made. Here we compare the attributes of the haptic sense—its ability to process information spatially and temporally—to those of vision and audition. This comparison demonstrates that it is an intermediary sense in terms of its capacities. An important feature of haptic perception is the way that information about the external world is acquired. We review the important role of different hand movements in extracting specific information about objects, showing these movements, known as exploratory procedures, to be both optimal and sufficient for the property of interest.

Haptic perception senses the physical properties of objects encountered in the environment, thus its focus is external rather than on internal tactile sensations. Since the hand is the primary structure used in tactual exploration of the external world, the study of haptic perception primarily involves studying the hand.[1] The distinction between haptic and tactile sensing, or between active and passive touch, is based on the active component of haptic sensing: the hand is voluntarily moved across a surface or manipulates an object to obtain specific information. For some properties, like perceiving the roughness of a surface or detecting minute surface irregularities, as in the finish on a car, it matters little whether the hand moves across a stationary surface (haptic sensing) or the surface is moved across the stationary fingers (tactile sensing). What is critical to the performance of such perceptual tasks is the relative motion between the fingers and the surface.[2] For other properties, such as the weight of an object, perception is facilitated by hand movements; for example, we are much more sensitive to weight when an object is lifted and jiggled up and down than when it rests passively on our outstretched hand.[3] The ability to make judgments about properties like weight or surface texture depends critically on two senses, touch and kinesthesia.

People can generally identify common objects haptically within a couple of seconds of contact. When we pick up an eraser, its material properties, such as surface texture or compliance (hardness/softness), are easy to perceive and play an important role in the recognition process. Think of the ease with which we can determine with our eyes closed whether it is a golf ball or an egg that has been placed in our hands. An object's geometric characteristics, such as its shape or volume, must generally be determined by following the contours of its shape, and so the perception of these attributes requires integration of haptic information over time. The haptic system is therefore much quicker at processing an object's material as compared to its geometric properties.

The interaction between the fingers and the surface of an object, when broken down into its basic elements, is revealed to be surprisingly complex. The finger pad becomes flatter as pressure is applied, and there is lateral movement of the skin as it is compressed. The surface topography of the fingertip with its ridges affects this interaction, and then frictional forces between the skin and surface being explored affect the movements generated as the finger scans over the object's surface. Finally, as the finger moves, small and complex vibrations occur that are transmitted as traveling waves across the skin surface and sensed by all types of mechanoreceptors in the skin. The elements of this complex process are illustrated in figure 5. Collectively, these interactions enable us to perceive a wealth of information about an object; they form the basis of haptic perceptual experience.

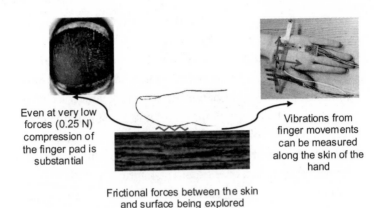

Even at very low forces (0.25 N) compression of the finger pad is substantial

Vibrations from finger movements can be measured along the skin of the hand

Frictional forces between the skin and surface being explored

Figure 5 Schematic representation of the interaction between the fingertip and an object. There is considerable compression of the finger pad, even at low contact forces, and as the finger moves across the object frictional forces are generated, which result in minute vibrations that travel along the skin surface.

Touch, Vision, and Audition

Vision and audition dominate our interactions with the world, so we tend to think of them as essential to our existence. However, as illustrated by the case of the individual with overwhelming sensory loss described in chapter 2, in the absence of our tactile and kinesthetic senses we are profoundly limited in our daily interactions. The tactile sense is considered to have a much lower bandwidth than either vision or audition, in that the amount of information it can process through time is lower than that of the other two senses. It has also been suggested that because the tactile modality is the first sensory system to develop—an embryo develops sensitivity to tactile stimulation by the eighth week of gestation—it may be more "primitive" than vision and audition. A number of metrics have been used to capture these differences among the sensory systems, and they reveal that in many ways touch is an intermediary sense.

Sensory systems are often compared in terms of their spatial and temporal resolution. The former refers to the spatial separation between stimuli that can be detected, for example, how close two probes can be on the skin for someone to perceive them as being distinct, a quantity known as the two-point threshold. On the fingertips, we can resolve a separation of about 1 mm, which places touch between vision and audition, visual spatial acuity being the best of the three senses. Spatial acuity can also be measured in terms of the feature size that can be detected. With no movement across the skin, small bumps around 0.2 mm can be detected, but when there is movement between the finger and the surface a feature as small as 13 nm (13 billionths of a meter) can be perceived on the fingertips.[4] The latter number indicates that human tactile perception extends to the nanoscale. Temporal resolution refers to the time difference required for two pulses delivered to the skin to be perceived as successive and not simultaneous. For touch, a temporal difference of 5 ms can be resolved, which is better than vision (25 ms) but worse than audition (0.01 ms). Touch is therefore an intermediary sensory system in that its spatial resolving power is poorer than that of vision but superior to that of audition, and its temporal resolving capacity is better than vision but inferior to audition.

Another metric sometimes used to compare sensory systems is their information processing capacity as defined in terms of the number of bits of information per second that can be processed. Since the fingertip is one of the most tactually acute areas of the body, its processing capacity has been compared to those of the ear and the eye. The eye is the most efficient; estimates are in the range of 10^6 bits per second (bps), next is the ear at 10^4 bps, and finally the finger at 10^2 bps.[5] Obviously these comparisons involve very different types of stimuli and tasks for each of the modalities, but the ranking does reflect the slower processing that characterizes the tactile/haptic sense.

Tactile Sensitivity and Acuity

Measures of spatial acuity on the skin include the two-point threshold described above; the smallest perceptible difference in the orientation of gratings (i.e., vertical versus horizontal) applied to the skin; and the minimum

Touch is therefore an intermediary sensory system in that its spatial resolving power is poorer than that of vision but superior to that of audition, and its temporal resolving capacity is better than vision but inferior to audition.

distance between two points of stimulation on the skin that is perceived to be different, known as point localization. For gratings, the smallest groove or ridge width at which the orientation is perceived defines the threshold; in the fingertips this is around 1 mm. Other measures of tactile acuity include the minimum level of pressure that can be applied to the skin and perceived, and the smallest feature that can be detected on an otherwise smooth surface. Pressure sensitivity is usually measured by indenting the skin with nylon monofilaments (similar to fishing line) of varying diameter until the filament buckles, and asking people whether they perceive the stimulus or not. The smaller the diameter, the lower the buckling force. Tactile pressure sensitivity is highest on the face, particularly around the mouth, followed by the torso and arms and hands, and lowest on the legs and feet. Gender influences pressure thresholds; women, on average, have lower thresholds (i.e., higher sensitivity) than men. However, it is possible that this sex difference in pressure sensitivity reflects finger size differences between men and women. The crucial variable that accounts for sex differences in tactile spatial acuity, as measured using the grating orientation task, is the size of the fingers. With decreasing finger size, tactile spatial perception improves, presumably due to the higher density of Merkel cells that cluster around the bases of sweat pores in the epidermis. The density of these mechanoreceptors, and of their associated

SA1 afferent units, is directly related to tactile spatial acuity. This means that women, who on average have smaller hands than men, have finer spatial acuity, and that a woman and a man with similar sized fingers will, on average, have similar acuity.[6]

Sensitivity can also be considered in terms of how people perceive changes in force or pressure that are well above the threshold level. For example, if I press on your fingertip with a force of 4 newtons, which is very perceptible, how much does that force have to change before you perceive a difference? A more concrete example would be if you were holding an apple in each hand and wanted to know how different the weights had to be for you to perceive them as being different. For force and weight, the change is around 6%, which means that an apple weighing 213 gm (7.5 oz) or 241 gm (8.5 oz) is perceived as being different in weight from an apple weighing 227 gm (8 oz). This difference, known as the differential threshold, is sometimes also referred to as the just noticeable difference (JND) or the *Weber fraction*. It has been calculated for a broad range of sensory attributes, and because it is dimensionless it can be used to compare the sensitivity of different senses.[1]

Fluctuation of pressure on the skin through time is referred to as tactile vibration. The sensitivity of the skin to vibratory movements has been measured as a function of the frequency of vibration. We can detect vibrations delivered perpendicular to the skin from about 0.5 Hz up to about 700 Hz, but are not equally sensitive to displacement of the skin across this range of frequencies. We are most sensitive to vibrations between 200 and 300 Hz and progressively less sensitive at lower and higher frequencies. The relation between vibration frequency and displacement of the skin is therefore a U-shaped threshold function. The four types of cutaneous mechanoreceptors found in glabrous skin (described in chapter 1) are sensitive to different frequencies of vibration within this range. At threshold levels of vibration, it is the type of mechanoreceptor most sensitive to a particular range of vibrotactile frequencies that determines the perceptual threshold: our capacity to detect very small vibrations between 200 and 300 Hz is due to the FA II receptors that respond to vibrations above 50 Hz, whereas FA I receptors respond to vibrations between 25 and 40 Hz.[7] For stronger vibrations well above these threshold levels, more than one type of mechanoreceptor typically responds to movements of the skin.

Perceiving Object Properties

Haptic exploration enables us to perceive both the geometric and material properties of objects. The former refers to features such as size, shape, orientation, and curvature, whereas the latter includes attributes such as surface texture, compliance, and thermal characteristics. The size and shape of objects that fit within the hand can be perceived on the basis of skin indentation and the pose of the fingers, because the hand can enclose the object. For larger objects, which must be sequentially explored by the hand to determine shape, kinesthetic cues also contribute to perception. This sequential nature of haptic exploration, in which a percept or image of an object's shape and size must be built up over time, imposes a load on memory processes that does not occur when shape is perceived visually and the object is seen in its entirety. This temporal aspect of sensory processing is one of the limitations of haptic perception of geometric properties. In addition, the process of perceiving geometric features haptically is subject to a number of biasing influences that systemically affect the veridicality of some percepts. For example, the haptically perceived length of a horizontal or vertical line is very similar to its physical length, but when its orientation becomes oblique its length is perceived less accurately. This effect also occurs when length is estimated visually. Several other visual illusions, such as the well-known Müller-Lyer illusion in which two lines of equal length are perceived to differ in length depending on whether the ends of the lines are enclosed by arrowheads or by fins, have also been shown to occur haptically. These illusions are described in more detail in chapter

4. The similarity in how the haptic and visual sensory systems process the geometric features of these stimuli indicates that their underlying perceptual processes are similar.

Surface Texture

Of all the material properties of objects, the one that has been subject to most research is *surface texture*, which can be further classified in terms of roughness, stickiness, slipperiness, and friction. As these terms suggest, texture is a multidimensional construct. Understanding how we perceive these various aspects of a surface is important in a number of areas, from the feel of consumer products that we hold, such as phones, handles, paper, and fabrics, to creating such textures artificially on flat screens. Of all the products purchased online, the one category that people report a significant need to touch prior to making an Internet-based purchase is clothing. The challenge of conveying the distinctive feel of different fabrics—for example, velvet, leather, or silk—to Internet purchasers is daunting, given the subtleties of the tactile sense that would have to be recreated for a display to be effective. The "feel" of fabrics is related to a number of features including their thickness, compressibility, and tensile properties. In addition, when we rub a fabric between our fingers, we notice how the surface moves as we create shear forces.

When people are asked to classify different textures in terms of their perceived tactile dissimilarity, three dimensions often emerge around which the various textures cluster. The first two dimensions are roughness/smoothness and hardness/softness, which are independent and robust. The third dimension is sticky/slippery, which is not as salient as the other two dimensions and not as reliably perceived across all individuals.[8] Stickiness refers to the sensations associated with making contact with adhesive tape or the resin on a pine tree, whereas slipperiness is experienced when the hand grasps a bar of soap that is wet. For this perceptual dimension, tangential forces are critical to perception.

Much of the research on surface texture has focused on roughness and understanding how judgments of roughness are made, including the influence of various parameters, such as the size and spacing of the elements that cover the surface, on these judgments. For relatively coarse textures like sandpaper, the perceived roughness relates to the gap between the elements comprising the surface, known as the spatial period, with larger spacing being perceived as rougher than textures with a smaller spacing. The perception of roughness for finer textures with much smaller spatial periods is based on the high-frequency vibrations generated by movement of the fingers over the surface.

Compliance

Compliance refers to how deformable a surface is when force is applied, so it is defined as the ratio between displacement, or the movement of the surface, and the forces applied. The inverse of compliance is stiffness. These terms define physically what we often refer to perceptually along a softness to hardness dimension. When we shop for food, we may be interested in evaluating the compliance of a piece of fruit or a soft cheese because that provides us with information about its ripeness or freshness. With deformable items like a mattress or cushion, we are capable of discriminating varying degrees of compliance or softness. The change in compliance that is perceived, that is the differential threshold, is around 22%, which is considerably larger than the 6% described for weight and force. A distinction is often made between objects that have compliant or continuously deformable surfaces, like a rubber ball, versus rigid surfaces such as a foot pedal in a car, which also moves with applied force. With soft or deformable objects, information from cutaneous mechanoreceptors is critical to evaluating compliance, whereas for rigid objects both tactile and kinesthetic (i.e., haptic) cues are essential. Understanding how people perceive the compliance of rigid objects is important to the design of all types of interfaces that move when forces are applied such as keyboards, control panels, and steering wheels. For most of these interfaces we want to know when we

have generated sufficient force to register a response—
this is called the breakaway force—and so tactile feedback
is often incorporated in the design of the display to signal
when the breakaway force is exceeded. For keyboards, the
force at which tactile feedback is presented ranges from
0.04 to 0.25 newtons.

Viscosity

A perceptual entity related to compliance is viscosity,
which physically is the ratio of force to velocity. We are
aware of viscous forces when we try to run in the ocean
or when we mix substances like cornstarch and water together to create a viscous mixture. One area where the perception of viscosity has historically been important is in
bread making and cheese and butter manufacture. These
activities require judgment of the rheological properties
(flow and deformation) of the materials as they are being manipulated. Although many of the processes associated with production of these foods have been automated,
some culinary procedures, like making sauces and knead-
ing dough for bread, still rely on discrimination of changes
in viscosity. With the advent of teleoperated robots another application for understanding the perception of viscosity has emerged, since these robots are often used in
viscous environments inside the human body or in the sea.
In these applications it is important to understand how
consistently human operators are able to discriminate

changes in viscosity as the robots are moving. Differential thresholds for viscosity have been estimated to be around 19%,[9] which is similar to the value reported for compliance (22%). This suggests that when force and movement cues have to be integrated, there is a loss in perceptual resolution since the differential thresholds for both force and limb movement are around 6% to 8%.

Exploratory Procedures

One of the interesting aspects of studying haptic sensing is determining what information we can acquire about an object as we manipulate it, and what types of hand movements are optimal for perceiving specific properties. For example, if people are asked to estimate the weight of a small melon, they will typically lift it up in the palm of the hand and move the hand up and down to perceive its weight. In contrast, if they are asked to judge how ripe the melon is, they will poke the surface of the fruit with one or two fingers to determine whether it is hard or soft. Such movements are referred to as exploratory procedures (EP) and are defined as a stereotyped pattern of manual exploration that is used when people are asked to judge a particular property.[10] These movements have been shown to be optimal, in that they provide the most precise information for extracting the property of interest. To perceive

If people are asked to estimate the weight of a small melon, they will typically lift it up in the palm of the hand and move the hand up and down to perceive its weight. In contrast, if they are asked to judge how ripe the melon is, they will poke the surface of the fruit with one or two fingers to determine whether it is hard or soft. Such movements are referred to as exploratory procedures.

temperature, for example, the EP of static contact is used in which a large area of the hand rests on an object without moving, maximizing the contact area between the skin and the object for heat flow between the two. This EP is optimal for perceiving temperature when compared to all other actions. The EPs that have been documented in laboratory studies are the movements that most people spontaneously use when asked to make a judgment about a specific property. The enclosure EP, in which the hand encloses the entire object to determine its global shape or volume, has been shown to be the most efficient way of judging the shape of an object. In contrast, to assess compliance (or hardness) the pressure EP is used to poke at the object and so determine its compressibility, whereas the lateral motion EP, in which a finger is moved back and forth across a surface, is used to perceive texture. Some of these EPs are illustrated in figure 6.

The EP that is executed to perceive a particular property, such as lateral motion to perceive texture, is in turn optimized to the specific situation in which it is being used. For example, people will vary the contact forces more when exploring smooth as compared to rough surfaces, and scan the surfaces more rapidly if they are asked to discriminate between them rather than just identify them. Similarly, when determining the compliance of an object, greater forces are used for rigid objects than compliant ones. In each of these situations, although the EP

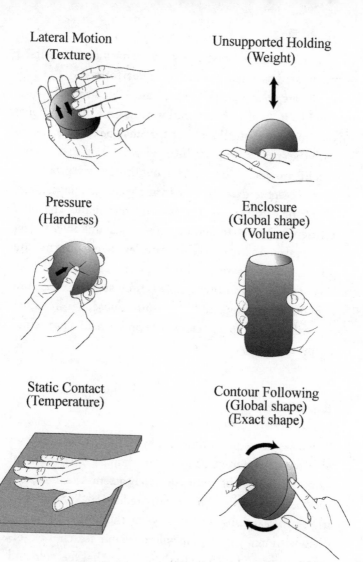

Lateral Motion
(Texture)

Unsupported Holding
(Weight)

Pressure
(Hardness)

Enclosure
(Global shape)
(Volume)

Static Contact
(Temperature)

Contour Following
(Global shape)
(Exact shape)

Figure 6 Exploratory procedures (EPs) and associated properties (in parentheses) that each EP is optimal at providing. From Jones and Lederman (2006)[1] and adapted from Lederman and Klatzky (1987)[10] with permission of Oxford University Press.

can be considered a stereotypical movement, it is itself modulated based on the task being performed.

There are costs and benefits associated with the various EPs in terms of exploration time and the additional information about the object that may be acquired along with the specific property being identified. The static contact EP provides information about the thermal properties of an object and at the same time yields incidental cues regarding its surface texture, shape, and volume. The most efficient approach to finding out about an unknown object is to grasp and lift it, as this provides coarse information about both its material and geometric properties. Three EPs are involved in this simple action: static contact, unsupported holding, and enclosure. Subsequently other EPs will be used to determine the presence of a particular property.

Haptic Search

When we attempt to find an object that is hidden from view we use different strategies to simplify the perceptual experience. The ability to selectively attend to salient features is an example of complexity reduction. For example, if we are searching for our keys at the bottom of a bag filled with many items, we will make our search more efficient by attending only to those items that feel cold and hard. In such situations we are focusing our attention on a specific feature or target that we hope will make our exploration time shorter. In studies of the visual system, features have been identified that are more distinctive than other properties and are said to stand out from others. This phenomenon is known as the *pop-out effect*.[11] An example from vision is the ease with which we can find a red circle among blue circles; it is seen immediately and we do not need to search through all the circles to find it. It is much harder, however, to find a slanted line among an array of vertical lines; in this case we need to search carefully through the visual scene to identify the line. That kind of search is a serial process in which items are explored one by one, whereas with features that pop out, the search occurs in parallel, and multiple items can be searched at once. The difference between these two strategies is reflected in the time taken to find the target—much longer response times are associated with the serial strategy, and also occur when the number of nontarget or distractor items increases.

When we identify haptic features that pop out we learn something about the processes used to recognize objects and discriminate between them. Haptic features that have been shown to pop out include roughness, coldness, and hardness. These features are not necessarily symmetric, that is, although a rough surface pops out in an array of smooth surfaces, a smooth surface does not pop out among rough surfaces. Other haptic features, such as the orientation of edges (vertical or horizontal), do not pop out, so individual elements in an array have to be serially processed to determine whether the selected feature is present. The high salience of material properties to haptic perception is reflected in the features that pop out during haptic search. These are also the features that people attend to when they are asked to judge the similarity of different objects using touch alone, because they are perceived quickly and efficiently. It is only when the same objects are visually perceived that geometric properties, like shape and size, become important in determining similarity.

HAPTIC ILLUSIONS

In the previous chapter, the properties of the haptic perceptual system were described. This chapter focuses on perceptions that are unexpected or surprising given the nature of the physical stimulus; such events are often referred to as illusions. Haptic illusions encompass a broad spectrum, ranging from distortions in the physical properties of objects to alterations in the representation of dimensions of parts of our body. There are also a number of haptic illusions that demonstrate how time and space affect perception, i.e., that tactile stimuli that occur close together in time are perceived to be closer spatially than they really are. Collectively, these illusions provide valuable insight into how people normally perceive and represent their environments. They also provide tools that enable us to enhance perception by compensating for missing components of the perceptual experience.

Illusions of Objects and Their Properties

Illusions provide an interesting window into the cognitive processes that people normally use to perceive and interpret events in the environment, since they reveal a surprising discrepancy between a physical stimulus and its corresponding percept. Although we are all familiar with a wide variety of visual illusions, we are less aware of illusions that affect the haptic sensory system. One illusion with which many people are acquainted is the size-weight illusion in which the larger of two objects with the same weight but different volumes is perceived to weigh less than the smaller object. This illusion occurs regardless of whether an object is perceived only haptically or both haptically and visually. A number of hypotheses have been proposed to account for it, including expectation theories that emphasize the role of previous experience in judging weight, particularly the expectation that larger objects will weigh more than smaller objects. This results in higher grip forces being applied to lift the larger object, which then accelerates more rapidly than anticipated, leading to the perception that it is light. However, this is not a complete explanation of the size-weight illusion, because even when the grip forces have adapted to the weight of the object after repeated lifting, the larger object is still perceived to weigh less than the smaller object.

Illusions provide an interesting window into the cognitive processes that people normally use to perceive and interpret events in the environment, since they reveal a surprising discrepancy between a physical stimulus and its corresponding percept.

The perceived weight of an object can be influenced by other properties besides volume, including temperature (as described in chapter 2), shape, density, and surface texture. The shape-weight illusion refers to the observation that objects that appear visually to be the smallest are often judged haptically to be the heaviest. Cubes are perceived visually to be larger than spheres of the same physical volume, and haptically they are perceived to be lighter. This illusion is closely related to the size-weight illusion in that shape and size, as sensed visually or haptically, influence perceived heaviness. The mass and volume of an object are related through its density, and density plays an important role in the determination of heaviness. Objects made from denser materials like brass are judged haptically to weigh less than those made from less dense materials such as wood, even though their mass is the same. The effect of surface texture on judgments of weight appears to be related to the grip forces required to lift an object. Objects with more slippery surfaces like those covered in satin are perceived to be heavier than those with rougher surfaces such as sandpaper, with the latter requiring smaller grip forces than the former to prevent them from slipping between the fingers. These weight illusions reveal the profound effect that the geometric and material properties of an object have on its perceived weight. As the volume, shape, material composition, contact surface, or temperature of an object changes, so too does the perception of its mass.[1] These illusions persist even when the grip forces used to lift an object have adapted to the actual weight of the object and not its expected weight. Several haptic weight illusions are summarized in figure 7.

A number of visual illusions involving distortions in the perceived size or linear extent of objects have been described, and these have been studied to see whether there are analogous illusions in the haptic modality. To the extent that such illusions are demonstrated haptically, they bring into question purely visual explanations of visual illusions and instead suggest a similarity in the underlying perceptual processing of such stimuli in the two senses. The Müller-Lyer and horizontal-vertical illusions (see figure 8), two well-studied visual illusions, also occur haptically. In the Müller-Lyer illusion a line bounded

Volume
Small objects heavier than large objects

Shape
Spheres heavier than cubes

Density
Wood heavier than brass

Surface texture
Objects with smooth surfaces heavier than those with rough surfaces

Temperature
Cold objects heavier than objects at skin temperature

Figure 7 Haptic weight illusions for five properties of objects. In each comparison, the objects have the same mass but differ with respect to the property specified.

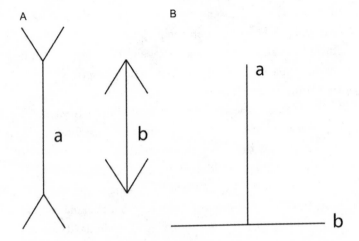

Figure 8 (A) The Müller-Lyer illusion, in which the line bounded by the fins (a) is perceived to be longer than the line of the same length bounded by arrowheads (b). (B) The horizontal-vertical illusion, in which the vertical line (a) is perceived to be longer than the horizontal line (b) of the same length. From Lederman and Jones (2011)[1] with permission of IEEE.

by arrowheads is perceived as being shorter than one bounded by fins, whereas in the horizontal-vertical illusion the length of a vertical segment is overestimated by about 1.2 times relative to a horizontal segment of the same length. The magnitude of the Müller-Lyer illusion is similar independent of whether length is perceived visually or haptically, and, interestingly, the presence of a strong Müller-Lyer and horizontal-vertical illusion in individuals who have been blind since birth indicates that visual imagery or experience is not necessary for these illusions to occur.[2] The presence of such illusions in both the visual and haptic modalities suggests that they are based on similar rules and representations of real-world properties. Other visual illusions such as the Delboeuf illusion, in which the perceived size of a circle changes if it is within a larger concentric circle as compared to standing alone, have not been demonstrated to occur haptically.

As these geometric illusions reveal, the haptic perception of space is systematically distorted with respect to physical reality and is non-Euclidian, as has also been noted for visually perceived space. The spatial distortions that arise haptically can be substantial, as shown by the errors that blindfolded individuals make when they are asked to rotate a rectangular bar in such a way that it is parallel to a reference bar. Huge systematic deviations are made that for some people can be up to 90° away from the reference orientation. These errors are persistent and robust, in that they remain even when people are provided with visual feedback about their errors after each trial.[3]

Illusions of Body Space

A number of tactile illusions involve distortions in the spatial processing of stimuli applied to the skin, such as the perceived location of stimulation or the perceived distance between tactile stimuli. These illusions generally result from interactions between the spatial and temporal properties of stimuli and demonstrate how time or the interval between stimuli affects their perceived representation on the body. Four tactile illusions demonstrate these interactions convincingly. They are illustrated schematically in figure 9.

If the time interval between stimuli presented on the skin is very short (100–300 ms) they are perceived to be closer together spatially than they really are, an illusion known as the *tau effect*. For example, when three tactile stimuli are delivered successively to the skin as shown in figure 9A, with the distance between the first two stimuli being twice that between the second and third, but the time interval between the second and third being double that between the first and second, the perceived distance between the stimuli is markedly affected. The distance between the second and third stimuli will be judged to be nearly twice as large as that between the first two stimuli. A related phenomenon occurs when a continuous tactile stimulus is applied to the skin, such as a probe moving along the forearm, and people have to judge the distance traveled. The perceived extent of the movement is strongly influenced by the velocity with which the probe moves across the skin, with a given distance being perceived as shorter by up to 50% when it moves at a faster (2500 mm/s) as compared to a slower (10 mm/s) velocity (see

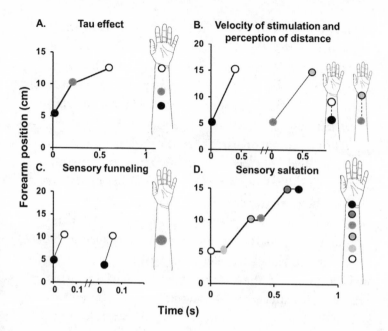

Figure 9 Four cutaneous tactile illusions. The temporal and spatial properties of the actual tactile stimuli presented on the skin are indicated in the graphs, and the illusory perceived sequences are shown on the forearm beside each plot. Figure from Lederman and Jones (2011)[1] adapted and redrawn from Goldreich (2007)[4] with permission of IEEE.

figure 9B). Interestingly, there is a range of velocities of movement across the skin (50–200 mm/s) for which the velocity has little effect on perceived distance. This range is within that used when people voluntarily scan textured surfaces and is optimal for detecting moving tactile stimuli on the skin.

The perception of distance on the skin is not uniform across the body, but varies with the particular area stimulated and the orientation of the stimuli presented. The distance between two points of stimulation is perceived to be greater in areas with higher spatial acuity, like the hand or face, as compared to those with lower acuity such as the leg or forearm, an effect referred to as *Weber's illusion*. At some locations on the body the orientation of the tactile stimulus also affects the perceived distance between two points of stimulation. On the thigh and arm, transverse distances are perceived to be greater than longitudinal distances of the same magnitude, but on the palm and abdomen, orientation does not affect perceived distance. These perceptual differences across the body may reflect variations in the geometry of receptive fields, or the contribution of anatomical landmarks such as joints to the perception of distance.

One illusion, in which tactile stimuli are perceived at locations in which there is no stimulation, is called the *sensory funneling illusion* (see figure 9C). When very brief tactile stimuli are presented simultaneously at several closely spaced locations on the skin they are often perceived as occurring at a single focal point at the center of the stimuli. This "funneling" of tactile inputs results in a stimulus that is perceived as more intense than the individual stimuli.

Illusions of movement have also been documented on the skin; these are of interest in designing tactile displays, because they provide a mechanism for creating directional cues such as tactile vectors. When a number of discrete mechanical taps are presented sequentially on the skin, they are usually perceived as a single stimulus moving across the skin and not as separate stimuli. This illusion of apparent movement, known as the *phi phenomenon*, is very compelling in terms of creating a continuous sense of movement. The optimal interval between stimuli for the illusion to occur varies directly with the duration of the stimuli; for 100-ms duration taps on the skin the optimal interval between the onset of each tap is 70 ms.

A related illusion that involves mislocalization of tactile stimulation on the skin is *sensory saltation* (see figure 9D). In this illusion a series of short tactile pulses delivered successively at three locations on the skin is perceived as a stimulus that is moving progressively across the skin "as if a tiny rabbit were hopping" in a smooth progression from the first to the third location.[5] This illusion was first described on the skin and came to be known as the *cutaneous rabbit*. It was subsequently shown to occur in vision and audition. As with all illusions there are optimal conditions for it to occur in terms of the number of taps (3–6), the time between stimuli (20–250 ms), and the area of skin across which the tactile stimuli are delivered. If the time interval between taps on the skin is 300 ms or more, then they are localized accurately and no illusion occurs. The region of skin over which the illusion occurs is known as the saltatory area and varies from a very small area on the fingers (2–3 cm^2) to a much larger area on the forearm (146 cm^2). What is particularly interesting about this illusion is that when people have to decide whether a stimulus was presented at a specific location on the body under the condition where a tap was actually delivered at the site and when saltatory stimuli were presented that evoked the illusion of a stimulus at the location, they cannot distinguish between the two situations. This means that the percepts evoked by real and illusory stimulation are phenomenologically very similar.

These tactile illusions of space and time demonstrate a fundamental principle of perceptual processing, namely that when two events occur in close spatial and temporal proximity the presence of one affects the perception of the other. In the context of tactile displays on the body, this means that physical distances may be perceived differently at different locations and that the perceived location of stimuli can change if the timing of tactile signals varies. However, such illusions also indicate a mechanism

These tactile illusions of space and time demonstrate a fundamental principle of perceptual processing, namely that when two events occur in close spatial and temporal proximity the presence of one affects the perception of the other.

for enhancing the display of tactile cues, in that with phenomena such as sensory saltation it is possible for users to perceive more inputs than are physically delivered.

Illusions of Body Representation

Tactile and kinesthetic signals provide us with information about the relative positions of our limbs in space and contribute to our body image, sometimes called the body schema. Several illusions involve distortions in our body schema that result in misperceptions of the size, length, or position of a limb. One such illusion is the *rubber hand illusion*, which refers to the perception that tactile sensations experienced on a person's unseen hand are referred to a visible artificial hand (a rubber hand), as if the visual input captures the tactile stimulation.[6] The illusion emerges when people see a rubber hand being stroked with a brush while their own unseen hand is simultaneously being stroked. A significant percentage of people report that they feel the touch of the brush on the rubber hand and not on their unseen hand. This effect can be quantified by asking people to indicate the perceived location of the unseen hand. Their hand is now displaced in the direction of the rubber hand. For the illusion to occur, the artificial hand must resemble the normal hand in size, shape, and orientation. It is also important for the tactile stimulation delivered at the two sites to be synchronous. This illusion has provided a tool to study the malleability of the body's image and has also been investigated to understand how visual, tactile, and proprioceptive signals are integrated.

Another illusion involving distortions in the representation of the body is the haptic vibration illusion. When vibration is applied to the skin overlying a tendon, illusory movements of the limb about which the vibrated muscle acts are induced. For example, if the biceps tendon that inserts into the radius bone in the forearm is vibrated, people perceive that the forearm is extending. When the triceps tendon on the back of the arm is vibrated, the arm is perceived to be flexing, even though it remains stationary. For the illusion to occur, the arm must be hidden from view and restrained so that it cannot move. To measure the perceived movement of the vibrated arm, investigators ask people to match the movements using their other arm, which is not vibrated. This illusion has been particularly important to our understanding of the sensory mechanisms involved in the perception of limb position and movement. It demonstrated the important role played by muscle spindle receptors in kinesthesia. These receptors are known to discharge intensely in response to vibration; the illusion occurs because the central nervous system and the brain interpret these signals as indicating that the muscle is being stretched.

A further interesting result from this research on the vibration illusion was the finding that if the muscle is actively stretched while vibration is applied, then people will often indicate that the vibrated limb is in an anatomically impossible position. For example, when vibration was applied to the muscles that flex the wrist while the wrist was maximally extended, people would say that it felt as if the hand was bent back toward the surface of the forearm.[7] This suggests that how our body is represented in the brain is not constrained by the anatomical limits of joint excursion and that the kinesthetic sense extrapolates from previous experience. Such plasticity in the body's representation is probably a very important aspect of our ability to use tools and other handheld devices as extensions of our limbs.

One final—and familiar—distortion in our perceptual representation of the body is the perceived alteration in the size of parts of our body when they are anesthetized. When an area of our mouth is anesthetized by a dentist we usually feel that our lips and lower face are swollen. These effects have been studied more systematically, by anesthetizing the digital nerves that innervate the thumb and measuring the change in its perceived size. The change is substantial—the thumb is perceived as 60% to 70% larger than normal. Interestingly, the lips are also perceived to increase in size (by 55%) when the thumb is anesthetized, even while the adjacent index finger's perceived size remains unchanged.[8] These changes reflect the sensory representation of the body in the cerebral cortex, where there are populations of neurons that respond to sensory signals from both the thumb and the lips; therefore it is not surprising that changes in inputs from one of these areas affects the sensory processing of information from both sites. The increase in perceived size has been interpreted as resulting from the unmasking of sensory inputs with anesthesia that are normally inhibited by feedback from the anesthetized area and the enlargement of the receptive fields of cortical neurons representing areas adjacent to the anesthetized region. These illusory effects are related to the extent of anesthesia and do not occur when a whole limb is anesthetized.

Haptic Aftereffects

An aftereffect refers to a change in the perception of a new stimulus after a period of prolonged stimulation with another (adaptation) stimulus. An example that we may all be familiar with, called the waterfall illusion, comes from vision. After looking at a waterfall for about a minute, we may notice that when we look at the rocks beside the waterfall they appear to be moving upward. This motion aftereffect captures one of the key features of an aftereffect, namely that the change is usually in the direction opposite to that of the initial stimulus. Aftereffects have been reported for a number of haptic stimuli, including curvature, shape, weight, roughness, and temperature; in each case the change in perceived magnitude is in the direction opposite to the adapting stimulus. The haptic curvature aftereffect refers to the finding that a flat surface feels concave after the fingers have repeatedly explored a convexly curved surface. This effect emerges rather quickly, after 2 to 10 seconds of manually exploring a surface, and can persist for up to a minute. These aftereffects are generally thought to be due to cortical processing (central) of sensory signals and not the result of overstimulation of peripheral receptors. This hypothesis is based on the finding that the aftereffect can be transferred to other digits on the hand not actively involved in the initial exploration of the stimulus, and for some stimuli, aftereffects can be demonstrated to occur on the opposite hand.

Applications of Haptic Illusions

As well as contributing to our understanding of haptic perception, tactile and haptic illusions can be utilized to manipulate human perception in efforts to enhance the display of information. Tactile displays can be designed to take advantage of the space-time illusions—such as mislocalization of tactile stimulation on the skin and the illusory perception of movement across the skin—to augment the signals delivered to a user. Illusory phenomena have been incorporated into virtual environments to enrich the perceptual experience. Haptic interfaces may be used to create mechanical signals that do not have counterparts in the real world, and their use can contribute to our understanding of the basic mechanisms involved in human haptic perception. For example, it had been assumed that haptic perception of curvature was determined by the surface geometry of the object in contact with the hand and not by the tangential forces created as the finger moved across the surface. However, it has been shown that when these two cues, which normally covary, are made to oppose one another in a haptic display, it is tangential force cues and not surface geometry that determine the perception of geometric features.[9]

With the development of virtual environments and augmented-reality systems, the need arises for haptic devices with performance capabilities that match the perceptual abilities of the user. Given the limitations of many haptic devices in terms of their bandwidth and the range of forces they can produce (see chapter 5), there has been an interest in using multisensory displays so that visual and/or auditory stimuli can be used to compensate for these shortcomings. For example, it has been shown that by changing the visual and auditory cues that are presented simultaneously with haptic cues signaling the stiffness of a virtual object, compelling haptic illusions can emerge. When the visual information presented on a screen or head-mounted display indicates that there is substantial deformation of a spring when it is compressed, even though the finger movements used to compress the virtual spring are not large, people attend more to the visual information in judging stiffness, disregarding the kinesthetic cues from the fingers.[10] A related effect has been demonstrated using auditory cues. When different impact sounds were presented as people tapped virtual objects using a haptic interface, auditory cues were found to influence the ranking of stiffness of the virtual objects, although the effect was not as robust as that found for visual cues.

Another category of phenomena that are used to create illusory haptic experiences is referred to as *pseudo-haptic feedback*. These effects use vision and the properties of human visual-haptic perception to create illusory experiences. Pseudo-haptic feedback has been used to simulate haptic properties such as the stiffness of virtual springs and the texture of a virtual object. It entails simulating haptic feedback by combining visual feedback in a synchronized way with the individual's movements or actions in the virtual environment, but distorting the visual feedback in some way to create the haptic illusion. For example, if the ratio between the movement of a user's hand on a computer mouse or other input device (C) and the visual displacement of an object on a screen (D) is altered so that the object appears to slow down or accelerate in a particular region that is visually distinct, then the change in the movement of the virtual object can be attributed by users to friction effects between the two virtual surfaces. The haptically perceived mass and the stiffness of virtual objects have also been manipulated by varying the C/D ratio to create these pseudo-haptic effects.[11]

TACTILE AND HAPTIC DISPLAYS

The sense of touch is engaged whenever we want to communicate using the skin. The signal may be as simple as an alert from a vibrating cell phone that indicates an incoming call or the shaking we experience when we drive across the rumble strips on the edge of a highway. In this chapter we examine the various types of tactile and haptic displays that have been developed, focusing on the technology used, the applications, and the challenges of developing wearable haptic systems. From the development of braille for the visually impaired in the early nineteenth century to the more recent advent of robotic systems that use haptic feedback to communicate remotely with their human controllers, success has depended on designing systems that are intuitive to use, do not encumber the user, and provide benefit that is not available visually or auditorily.

From the development of braille for the visually impaired in the early nineteenth century to the more recent advent of robotic systems that use haptic feedback to communicate remotely with their human controllers, success has depended on designing systems that are intuitive to use, do not encumber the user, and provide benefit that is not available visually or auditorily.

Most human communication involves vision and audition, but touch is also an essential component of human interactions. Studies of the impaired growth and development of infants deprived of physical contact in understaffed orphanages, and of the reduced incidence of respiratory problems and infections in preterm infants who have skin-to-skin contact with their mothers, provide evidence of the importance of tactile stimulation to normal human development. Looking beyond these developmental aspects of touch, the interest in using the skin as a medium of communication dates to the early nineteenth century and the development of braille. At that time, the focus was on substituting tactile information for the lost or absent visual input. More recently, with the proliferation of visual and auditory interfaces used to interact with computers and mobile devices—and the information overload experienced by users—there has been renewed interest in using the sense of touch to communicate. Advantages of the tactile mode include the large area of skin available to present information, the effectiveness of tactile stimulation in capturing our attention, and the underutilization of this channel of communication.[1] Tactile communication is also private, an important consideration in many contexts. A variety of tactile displays have been developed to support different communication requirements, ranging from wearable vibrotactile displays that assist in navigation to refreshable braille displays that enable those with visual impairments to read text.

Tactile Display Technology

Tactile displays can be divided into three broad classes based on the type of input delivered to the skin: vibration, static pressure in which the skin is indented, and lateral or tangential skin stretch. The technology used to deliver these inputs varies, from the electromagnetic motors used to present vibrotactile cues in wearable displays, to the dense arrays of electrodes that comprise electrotactile displays. Vibrotactile displays, which stimulate the skin mechanically using a motor that converts electrical energy into mechanical displacement either of the whole motor or of a contactor pad or pin on the skin, usually operate at frequencies ranging from 10 to 500 Hz. The motors are generally electromagnetic or piezoelectric; electromagnetic motors are more widely used because of their size, availability, low cost, and low power requirements. For these reasons they are frequently used in cell phones, pagers, and game controllers, where the vibration signal is presented to the whole hand.

Electrotactile displays create tactile sensations by passing brief (50–100 μs), constant current pulses (0.1–10 mA) through surface electrode arrays attached to the skin. The intensity of electrical stimulation is modulated by way of variation in the pulse duration and current amplitude. In contrast to displays that mechanically stimulate the skin and so activate the underlying mechanoreceptors, electrotactile displays activate nerve fibers that convey information from all types of cutaneous receptors. The sensations evoked by electrotactile stimulation are more diffuse and are described as tingling, pressure, and sometimes a sharp pain, depending on the stimulating current and waveform and on the degree of hydration of the skin. The stimuli delivered by electrotactile displays are highly dependent on the contact between the skin and the electrode array. These displays do not contain any moving parts and so are relatively simple to control and maintain. They are usually compact and have lower power requirements than electromechanical actuators—an important consideration for wearable displays powered by batteries.

Two issues with electrotactile displays that have limited their widespread adoption are (1) that they have a rather limited dynamic range (10 dB) and (2) that a moist environment is required to ensure efficient transfer of current from the electrodes to the skin. Dynamic range refers to the ratio of the largest to the smallest intensity of the stimulus, measured in decibels (dB). For electrotactile displays it is defined in terms of the current at which pain is first experienced (largest intensity) and the current at which an absolute threshold is measured (the smallest perceivable intensity). A small dynamic range means that the difference between threshold levels of stimulation and the onset of pain is rather small, and so stimulation current must be tightly controlled to avoid painful sensations. (In contrast, vibrotactile displays that mechanically stimulate the skin have been estimated to have a dynamic range of around 40 dB.) One electrotactile display that has been commercialized for use by the blind is the BrainPort (Wicab, Inc.), which converts the pixels in images captured by a camera mounted on glasses into strong or weaker electrical impulses felt by the tongue (this system is described in more detail in chapter 6). The electrotactile display itself is attached to the roof of the mouth and when the tongue makes contact a spatial pattern is experienced.[2]

Static displays in which the skin is indented have generally been developed to present cues such as braille to the fingertips. These devices require an array of pins, each of which must be individually controlled to represent the six cells that comprise braille characters. This necessitates the use of an actuator technology that is compact so that each of the pins can make contact with the finger pad. For this application, piezoelectric actuators have been used in a number of devices.

A different technology, based on airborne ultrasound, has recently been explored to deliver pressure sensations to the skin using a noncontact display. By spatially modulating the radiation pressure of the airborne ultrasound generated by an array of small speakers, the system can create 3D images that do not depend on vibrotactile stimulation.[3] The display can present virtual shapes that are manually explored and feel like gentle pressure on the skin. This technology is being examined for use as a non-contact tactile display in cars.

The final type of mechanical input that has been used in tactile displays is skin stretch. As the name indicates, this entails stretching the skin tangentially, in contrast to the indentation normal to the skin surface that is typical of static displays. Skin stretch displays can present directional cues to the fingers or palm by stretching the skin in the direction of the proposed movement (e.g., to the left or right) and orientation cues can be provided by rotational skin stretch.[4] These displays have been developed commercially and are being integrated into game controllers for virtual reality gaming. In this context, skin stretch is used to mimic the shear forces and friction that are experienced in the real world when holding an object. Lateral force cues can also be used to provide haptic cues regarding curvature, a geometric property.

Applications of Tactile Displays

Gaming

Vibrotactile displays have been part of arcade video games and video game consoles for many years. Beginning in the 1970s and 1980s with Sega's Moto-Cross and Atari's TX-1 arcade games, vibrotactile feedback became essential to the user experience, conveying the physical interactions of a vehicle with the road, such as going over bumps, or with other vehicles when they crashed. By the 1990s vibrotactile feedback was an integral part of most driving simulators. As game consoles developed and mass-market systems expanded, environmental effects such as the firing of weapons or the success of a knockout were conveyed tactually to the hand. Although the technology now used in these systems enables the delivery of more complex and more intense vibrotactile feedback to the user, due to better and more versatile motors, the effects generated are still relatively simple. In addition to vibrotactile feedback, handheld controllers have been used that also present shear forces that stretch or rotate the skin. By varying the rate and direction of skin stretch, these devices can convey mechanical impact, wrenching motions, and inertial cues. In the gaming market, there continues to be a demand for devices that can convey a range of realistic physical interactions haptically.

Consumer Electronic Devices

In many consumer devices like smartphones or watches, the simplest form of tactile communication is the notification that occurs when the phone or watch vibrates to indicate an incoming call or upcoming appointment. This feature has been implemented in many devices and simply alerts the user to an event. Vibration cues are used in preference to pressure cues because we are better at detecting a dynamic or varying signal than static stimulation. On many devices it is possible to customize the vibration, enabling different vibration patterns to be associated with different functions. This increases the information content of the vibration signal but requires that the user pay greater attention to the incoming message. For example, the rhythm of a vibration can be changed by making the time between pulses shorter to convey a sense of urgency; this may be used to signal an upcoming appointment or an important text message.

Vehicles

Tactile driver assistance systems have been implemented in a number of trucks and cars to alert drivers who may be fatigued or distracted and unaware of impending danger. Tactile feedback has also been incorporated in infotainment devices in some vehicles to avoid the distraction associated with drivers taking their eyes off the road to view a screen. Tactile warning systems have a number of advantages in comparison to auditory warning signals: they are relatively unaffected by the level of auditory background noise, are delivered specifically to the driver and not to other passengers in the vehicle, and can be more effective than auditory signals at presenting directional cues to a driver. These tactile warning systems represent one application where the displays have reached such a level of maturity and demonstrated effectiveness as to warrant their widespread use. It is estimated that by 2020 all new cars will be fitted with some type of tactile display that alerts the driver. In contrast to some of the other domains where tactile displays are being introduced, in this application automobile manufacturers are requiring that the system be easy and intuitive for the driver to use, and require no specific training.

Two systems have dominated the application of tactile displays in vehicles, *collision avoidance systems* that alert the drowsy or distracted driver about the proximity of obstacles or other vehicles, and *lane departure warning systems*.[5] Tactile displays have been mounted in the driver's seat, seat belt, steering wheel, and foot pedal. The collision avoidance systems are designed to help the driver avoid front-to-rear-end collisions, which are among the most common types of vehicular accident. Although tactile warning signals are an effective means of alerting drivers about a potential impending collision, auditory cues, such as the sound of a car horn, have been shown to be more effective than tactile signals in this context. Lane departure warning systems have been implemented in the car seat, in some vehicles, and in the steering wheel in others. The side of the seat or steering wheel that vibrates indicates the direction of the lane departure. Both systems have been shown to be effective as warning indicators,

These tactile warning systems represent one application where the displays have reached such a level of maturity and demonstrated effectiveness as to warrant their widespread use. It is estimated that by 2020 all new cars will be fitted with some type of tactile display that alerts the driver.

and both resulted in faster reaction times than an auditory alert. Most of these systems have used abrupt-onset tactile warning signals. There is an interest in determining whether graded tactile warning signals may be as effective and less annoying, but possibly also less immediate in that drivers may take longer to respond.

Navigation

Another successful application of tactile display technology uses the location of stimulation on the body as a spatial cue about the direction of intended movement. The displays can be worn as belts, vests, or wristbands and have been developed to assist in navigation in an unfamiliar environment or when visibility is limited, such as at night. Much of the work so far on developing wearable tactile displays for navigation has been focused on understanding what information is needed by the user, rather than on developing commercially available systems. In this application, tactile displays typically comprise an array of 8 to 16 small vibrating motors that is mounted on a belt or a vest and worn over clothing on the body. Such a display conveys information to the wearer by varying the temporal and spatial sequence of activation of the motors. If the individual needs to move to the left or right, a motor on the corresponding side of the body is activated, or several motors may be activated in a spatial sequence that goes toward the left or the right (see figure 10).

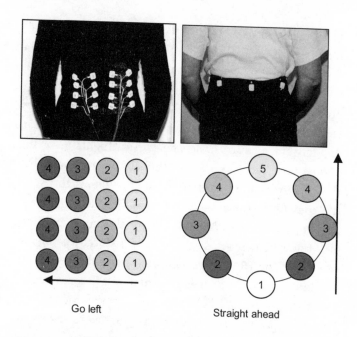

Go left   Straight ahead

Figure 10 Vibrotactile displays worn on the back or waist that can be used to convey directional cues to users to assist in navigation. The sequence of motor activation for the directions indicated is shown in the lower icons. For the display on the left four motors are activated concurrently as the sequence moves across the back.

One such system, known as the Tactile Situation Awareness System (TSAS), was developed in the 1990s and tested as a navigation aid for Navy pilots. In one prototype, 22 pneumatic actuators were mounted in a vest worn under a flight suit and activated using oscillatory compressed air that forced the membrane on the actuator to vibrate at 50 Hz.[6] The TSAS was designed to provide pilots with information about the velocity direction of the helicopter and its velocity vector magnitude so that they could control hovering (motionless flight over a reference point). A computer connected to the altitude and velocity sensors on the helicopter translated the craft's position and movement into tactile signals. Test flights involving pilots flying helicopters were conducted and showed that the aircraft could be controlled using predominantly tactile cues (with visual cues significantly reduced). There was improved control of the aircraft during these complex flight maneuvers, and the TSAS reduced the pilots' workload. The TSAS and other torso-based displays demonstrate that tactile signals presented to the torso can be used to control a vehicle and maintain spatial orientation. This work has also shown that it is possible for pilots to perceive vibrotactile signals presented on the skin in the presence of vehicle vibrations and other forms of environmental noise.

The torso provides a natural site for attaching such a display, since it seems to be very intuitive to perceive an external direction emanating from a single point or spatial sequence of stimulation on the body. With a two-dimensional array of motors, it is possible to create simple vibrotactile patterns that users can readily learn and respond to, simply by varying the number, location, and duration of motors activated, as illustrated in figure

10. These patterns can provide additional cues related to navigation—for example, instructing someone to stop by simultaneously activating all the motors in the display, or to slow down by changing the time interval between activation of different motors. Such displays have appeal for use in hazardous environments where visual and auditory communication may be limited by environmental factors such as low illumination or high noise levels. For example, firefighters and special operations forces can use tactile displays for communicating simple commands and instructions.

Haptic Displays

A distinction is generally made between tactile displays that provide vibrotactile, electrotactile, or static tactile inputs only to the skin, and haptic displays that engage both the tactile and kinesthetic senses. Haptic interfaces enable a user to make contact with an object in a real or virtual environment, to feel properties such as its weight or stiffness, and to manipulate the object directly.[7] The interface may be a stylus or a glove that provides a means of interacting with computer-generated virtual environments; or an exoskeleton worn on the hand and arm that is used to remotely control a robotic arm and applies forces to the user's hand and arm. Unlike tactile displays, haptic displays permit interaction between the user and the robot or virtual world that is bidirectional, with position (or force) being measured at the user's hand and forces (or position) fed back to the user. This means that in a surgical virtual environment, such as may be used for training residents in various surgical procedures, when the haptic device pushes against a tumor in the virtual lungs, for example, the user feels the resistive force, and its magnitude is proportional to the distance the tumor is compressed. Devices that sense position and feed back forces are known as impedance-type haptic devices, whereas those that sense force and feed back position are called admittance-type haptic devices. The former are more common because they are simpler to design and less expensive to produce.

Like tactile displays, haptic displays use a range of actuator technologies, but for these devices the motors are larger and more powerful, with a high control bandwidth, since they need to generate substantial forces. Brushed DC motors are the most commonly used actuator technology, but displays built with pneumatic, hydraulic, shape-memory alloy, and magnetorheological (MR) fluid motors have also been developed. Haptic displays have position and force sensors to measure the position of the user's hand or end effector and to control the forces fed back to the user. Power sources and electronics drive the actuators, operate the sensors, and enable communication between the haptic device and a computer.

The features that are generally optimized in a haptic device include low inertial mass, low friction, high force bandwidth and dynamic range, backdrivability, and a large workspace. Low mass is extremely important for any device worn on the body, such as an exoskeleton, so that user fatigue is minimized. For actuators, this entails maximizing the power-to-mass and power-to-volume ratios so the device is light, compact, and powerful. Although still important for performance, low mass is less of an issue with devices that are grounded on a desk or the floor, since those surfaces support the weight of the device. Low friction is critical to the capacity of a device to simulate fine textures and small interaction forces, which may be masked at higher levels of friction. A high force bandwidth ensures that the forces experienced by the user are perceived as stable and smooth. The concept of backdrivability refers to the characteristic of an actuator in a haptic device that can follow the user's movements rapidly and without imposing constraints on the motion. The workspace is the volume over which the device can operate; larger volumes allow more natural movements by the user and entail fewer constraints on the motion.

The workspace of different haptic devices in part reflects their domains of application. For those developed to control robots, in which there may be a one-to-one mapping of movements of the human operator's arm to those of a robotic arm, relatively large workspaces involving movement of the whole arm may be required. If the haptic device is worn on a freely moving hand then the mass of the device is an important consideration. Smaller haptic devices that have lower mass and better performance are more likely to be used in virtual or simulated environments such as computer-aided design (CAD) or medical training scenarios. In the latter case most of the interaction with the virtual environment is point-based; that is, the user feels only the contact between the end point of the stylus or thimble and the remote environment.

Haptic Rendering

The process by which the interaction forces are calculated between a haptic interface represented in a virtual environment and the virtual objects present in that environment is called *haptic rendering*. Software implements haptic rendering algorithms that define the properties of the haptic virtual objects and virtual environment, and the device hardware and software comprise the haptic interface. The haptic feedback provided to the user in a virtual environment depends on the simulation modeling and control algorithms. The virtual representation of the haptic device is the avatar, and this is what is used to interact physically with the environment. The avatar will vary with different applications, from a surgical tool like a scalpel in medical simulations, to a glove in an entertainment application. Physical modeling focuses on determining the dynamic behavior of virtual objects, principally on the basis of Newtonian physical laws. This means that when two solid objects collide they do not penetrate each other, and that when a compliant virtual object is grasped there is deformation of the grasping surface consistent with its physical properties. It is important to the user of a haptic device that when a tool is moved across a rough surface the texture is rendered in a manner that feels natural and consistent with the visual image of the object. These physical interactions have to be relayed in a realistic manner to the user, both visually and haptically, so that the collision response at the moment of impact indicates to the user whether the objects deform, bounce, and so on.

Haptic rendering algorithms detect collisions between objects and avatars in the virtual environment, and force response algorithms compute the interaction forces between the avatar and virtual objects when a collision occurs. These forces are then transmitted to the user through the haptic device. For this feedback loop to be effective, a high bandwidth and low simulation latency are required. It is often suggested that the minimum sampling rate required for haptic interfaces to be effective is 1 kHz. Computationally, physical rendering is more expensive than modeling based on geometry. With the latter, the interaction of the haptic device with the virtual environment is rendered based on geometric manipulations and so focuses more on the visual representation rather than the underlying mechanical deformation.

Applications

Haptic devices were first developed in the 1950s and 1960s as part of teleoperated (remote) robotic systems. In these systems, an operator controlled a master arm that transmitted commands to a slave device working at some distance in a hazardous environment, for example, in a nuclear power plant, in space, or underwater. There were actuators in the master arm that received signals from sensors in the slave device and enabled the operator to feel the remote forces. One of the earliest haptic display projects, the GROPE, was focused on developing a visual-haptic display that could present information regarding the force fields of interacting protein molecules in a virtual environment. This ambitious effort lasted more than 20 years, and by 1990, with the availability of much faster computer hardware, the goal of creating a three-dimensional molecular docking simulation was achieved.[8]

Haptic devices became commercially available in the late 1980s and 1990s in a variety of configurations, from joysticks (e.g., Impulse Engine, Immersion Corporation)[9] to thimble gimbals and styluses (e.g., Phantom, 3D

Systems, formerly Sensable Technologies)[10] and exoskeletons (e.g., CyberGrasp, CyberGlove Systems). These force and torque feedback devices vary with respect to their linkage mechanisms, workspace, degrees of freedom of movement, bandwidth, peak force, and stiffness. Serial linkage devices, such as the Geomagic Touch (formerly Phantom) or Virtuose (from Haption),[11] usually have a larger workspace than parallel-link devices (e.g., Delta or Omega devices from Force Dimension),[12] whereas the latter can generally output higher forces. Some of these devices are depicted in figure 11. As with most electromechanical systems, there are issues with many haptic displays related to

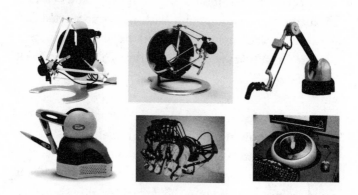

Figure 11   Commercially available haptic devices. Top row: delta.6 (courtesy of Force Dimension), omega.7 (courtesy of Force Dimension), Virtuose 6D desktop (courtesy of Haption). Bottom row: Geomagic Touch (courtesy of 3D Systems), CyberGrasp (courtesy of CyberGlove Systems) and Maglev (courtesy of Butterfly Haptics).

friction, caused by bearings or cables, backlash and limited bandwidth.

Haptic devices based on magnetic levitation became commercially available around 2007 with the Maglev (Butterfly Haptics).[13] These devices overcome some of the limitations of electromagnetic devices, such as mechanical backlash and static friction, since they have a much higher bandwidth and very high position resolution. With the Maglev, the user grasps a handle that has six degrees of freedom and is attached to a "flotor" that is levitated by strong magnetic fields. Forces and torques are fed back to the handle, which can move within the rather limited workspace of a 24-mm-diameter sphere. The features of this device are compatible with its use in virtual surgery or virtual dentistry where both hard and soft tissues need to be simulated and the workspace is generally of limited size.

Medicine With the emergence of virtual reality technology for training in medicine came the need for haptic devices to be developed as part of the simulation environment, so that the mechanical properties of tissues could be incorporated into the virtual training scenarios. These simulators reduced the need for practicing surgical procedures on animals and cadavers, with the associated costs and safety issues. They also provide the opportunity to measure the performance of trainees and to develop metrics that can be used to specify a level of competence required before operating on a real patient. As the surgeon-author Richard Selzer noted in his book *Letters to a Young Doctor*, a "facility in knot tying is gained only by tying ten thousand of them."[14]

Haptic interaction is an essential element of many medical and surgical procedures, from palpation, in which the physician manually probes body organs or tissues, to suturing, during which a needle is inserted through tissue and knots are tied, or maneuvering endoscopic instruments through the body to reach a target location. In order for these interactions to be simulated effectively, the haptic sense must be engaged. The haptic displays developed for medical and surgical applications are usually tool-based displays incorporating end effectors that can simulate needle insertion for suturing or the maneuvering of an endoscope in a body cavity. Much of this work is still at the research stage and uses commercially available haptic devices. Simulators with haptic feedback have been built for dental training that incorporate procedures such as drilling and the identification and removal of dental caries.

Haptic gloves have been used in a number of clinical research settings to assist in retraining hand use for patients who have lost the ability to move or feel sensations from the hand as a result of stroke or other neurological disorder. These gloves can be used both to measure the movements made and to deliver augmented feedback to the user that can assist in rehabilitation. When the gloves are coupled to a virtual environment it becomes possible to create games and provide goals for performance that encourage patients to practice and—importantly—allow the individual to exercise in the home environment rather than in a hospital or clinic.

Entertainment With the development of virtual environments that were used as racing and flight simulators came steering wheels or joysticks that included haptic displays. These devices can produce resistive forces associated with traveling through mud or water, and vehicle-based centrifugal forces that simulate the effects of high-speed turns. Because of their size, power, and performance requirements, such systems are expensive and so are generally reserved for training or for use in large-scale facilities rather than personal use at home. For the latter applications, as detailed earlier in this chapter, most systems that have been developed present tactile feedback to the hand via vibrating motors or actuators that stretch the skin to create tangential forces.

TACTILE COMMUNICATION SYSTEMS

The sense of touch is a very natural candidate for helping those whose daily lives are limited by impairments involving other sensory modalities. This chapter considers how the sense of touch has been used to compensate for deficiencies in the visual, auditory, and vestibular systems. For the visually impaired, tactile displays have been developed to assist in "reading" text, such as braille, and to aid mobility by providing cues regarding obstacles in the immediate environment. Tactile displays that compensate for impaired hearing range from systems that translate environmental sounds into patterns of vibration on the skin, to devices that assist with learning speech reading. For all these applications, there is interest in developing tactile vocabularies to enhance the capacity for communication. The chapter will conclude with an overview of this work and the challenges associated with creating tactile words, known as tactons.

Sensory Substitution

Sensory substitution refers to the use of one sensory modality to substitute or compensate for loss of function, due to disease or trauma, in another sensory system. The sense of touch has been used as a substitute for vision and for audition because it can process both spatial and temporal information. The visual system is extremely effective at processing spatial information about the external environment, and so is our primary means of encoding distance, the size and shape of an object, or the direction in which something is moving. It is much less capable of making discriminations based on the time between events, known as temporal discrimination, for which the auditory system is exquisitely tuned, as reflected in our capacity to process speech and music. The ear is capable of distinguishing between two auditory clicks separated by as little as 1.8 ms, whereas two mechanical pulses delivered to the skin on the fingertip must be spaced 10 ms apart for the two pulses to be perceived as distinct.

Tactile Systems for the Visually Impaired

One of the oldest and most successful sensory substitution systems is braille, in which information (text) usually processed visually is instead acquired tactually through the fingertips. This tactile alphabet was developed in the early nineteenth century by Louis Braille, who lost his sight at the age of three after an accidental injury. The alphabet was based on a 2-by-3 matrix of raised dots, which was a simplified version of a 12-dot (2-by-6) alphabet that had been devised by a French army captain, Charles Barbier. His goal was to create a communication system that could be used by soldiers to read battle commands silently and in the absence of light, and was known as night writing. Louis Braille recognized two limitations of Barbier's alphabet: first, that the 12-dot matrix was too large, in that it required moving the fingertip across each pattern to decode it; and second, that it made more sense for each pattern to represent an individual letter rather than a sound (phoneme). By trial and error, Louis Braille developed a new alphabet for the blind that was logically constructed and could be used to transcribe any language. The present physical structure of each braille cell, with an interdot distance of 2.28 mm and a dot height that varies from 0.38 mm to 0.51 mm, depending on the printing materials, is very close to that proposed originally by Braille. These values have been shown to be optimal in terms of reading speed and accuracy.

Braille is a static display that visually impaired individuals "read" by moving the fingers across the line of text from left to right. Most readers prefer to use two hands rather than one, and bimanual reading is faster than unimanual reading, in part because it enables faster transitions between lines of text. One finger can be used for

By trial and error, Louis Braille developed a new alphabet for the blind that was logically constructed and could be used to transcribe any language. The present physical structure of each braille cell, with an interdot distance of 2.28 mm and a dot height that varies from 0.38 mm to 0.51 mm, depending on the printing materials, is very close to that proposed originally by Braille.

reading the textural information while the other finger processes more spatial information, such as interletter or interword spaces, and ensures type alignment. Proficient readers can read braille at around two words per second, which is about one half to one quarter the rate typically achieved for visual reading of English text (five words per second).[1] At this reading rate, a single index finger scans 100 to 300 separate braille cells in 60 seconds. More recently, refreshable electronic braille displays have been developed to enable visually impaired individuals to read text on any computer or mobile device. In these displays movable pins generate braille, and between 40 and 80 characters from the screen can be displayed, depending on the particular electronic braille model.

Over the years, a number of devices have been developed to present visual information to those with visual impairments. These systems typically used a camera to capture an image that was then processed by a computer into a two-dimensional tactile pattern, presented on the body using an array of vibrating motors. One such device was the Optacon (*op*tical to *tac*tile *con*verter), developed by Bliss in the 1960s, which converted printed letters into a spatially distributed vibrotactile pattern on the fingertip using a 24-by-6 array of pins.[2] A small handheld camera was used to scan the text with one hand, and an image roughly the size of the print letter was felt moving across the tactile display under the fingertip on the other hand.

Although reading rates were much slower than those achieved with braille—highly proficient users could read 60 to 80 words per minute—the Optacon was one of the first devices to provide visually impaired people with immediate access to text and graphics. Between 1970 and 1990 more than 15,000 Optacon devices were sold, but beginning in the mid-1990s this technology was replaced by page scanners with optical character recognition, which were less expensive and easier for blind people to learn to use. Those systems have now been largely replaced by speech synthesizers and screen reader technologies.

A much larger system designed to enhance mobility for people with visual impairments was developed by Paul Bach-y-Rita and his colleagues in the late 1960s. This system attempted to address the critical mobility problems faced by visually impaired individuals who need to detect, identify, and avoid obstacles as they move. It was known as the Tactile Vision Substitution System (TVSS) and used 400 vibration motors mounted on a chair to display images on the back that were captured by a CCD camera. Users could recognize simple shapes and the orientation of lines, and with extensive training they could identify tactile images of common objects. In the 1960s and 70s TVSS devices had limited spatial resolution and a low dynamic range. They were cumbersome, noisy, and power hungry and so had limited impact on improving the mobility of visually impaired individuals.[3] More recently developed systems use small cameras mounted on remotely controlled mobile robots or on the user's head to acquire information about the environment. The video feed from the camera is translated into signals that activate electrodes or motors mounted on the body. One such device, the BrainPort, converts the pixels in the images captured by the camera into strong or weaker electrical impulses felt by the tongue. This electrotactile display is attached to the roof of the mouth and is designed to assist blind people using a cane or a guide dog to help them determine the location, position, size, and shape of objects as they navigate.[4] An interesting aspect of these TVSSs is that after intensive training users come to perceive the objects that are being displayed as tactile stimuli on their back, tongue, or abdomen as situated some distance in front of them, a phenomenon known as *distal attribution*.

Tactile Systems for the Hearing Impaired
The development of tactile communication systems for those with hearing impairments dates to the 1930s when electrical stimulation of the skin was used to teach deaf-mute children to understand speech. More recently, tactile-auditory substitution aids have focused on two primary applications. First, the translation of environmental sounds detected by a microphone—like knocking on a door, or a baby's crying—into time-varying patterns of vibration on the skin. For example, footsteps may be

represented by a series of brief low-frequency pulses and a fire alarm as a continuous high-frequency vibration on a device worn around the wrist. Second, tactile devices have been designed to convey aspects of the acoustic speech signal, to assist with speech reading for people with hearing impairments or to improve the clarity of deaf children's speech. Variables such as the intensity and duration of syllables and phonemes may be encoded by one vibrator operating at a low frequency while specific phonetic features of consonants, such as fricatives (e.g., *f* and *th*) and unvoiced stops (e.g., *p* and *t*), are displayed using higher-frequency vibration at another location on the arm. These systems have been demonstrated to be useful in learning speech reading, although it is clear that systematic and long-term training is required for hearing-impaired individuals to achieve maximal benefits from the tactile aids.[5] This is due to the arbitrary nature of the relation between the acoustic information and the associated tactile cue, which must be explicitly learned.

Tactile Systems for the Deaf-Blind

One sensory substitution system that makes use of the exquisite capabilities of the haptic sense was developed for people who are deaf and blind, and is known as Tadoma. The user's hand is placed on the speaker's face and monitors the articulatory motions associated with the speech-production process as shown in figure 12. This method is

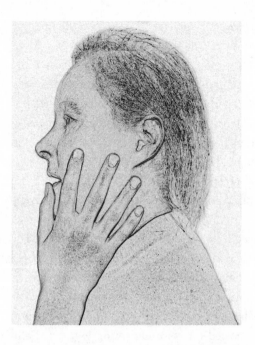

Figure 12 Position of the hand used in the Tadoma method to understand speech, with the fingers sensing movements of the lips and jaw, vibration of the larynx, and airflow from the mouth. From Jones and Lederman (2006)[9] with permission of Oxford University Press.

believed to have been developed by a Norwegian teacher in the 1890s and was first used in the United States in the 1920s, by Sophia Alcorn, to teach two deaf and blind children, Tad Chapman and Oma Simpson, after whom the technique is named. The physical characteristics of speech production that are sensed by the Tadoma reader include airflow, which is felt by the thumb on the lips, the laryngeal vibration on the speaker's neck, and the in-out and up-down movements of the lips and jaw, which are detected by the fingers. The haptic information provided to the hand of the Tadoma reader is multidimensional, but it is of sufficient resolution that very subtle changes in airflow or lip movement can be detected and interpreted in terms of a linguistic element.

The speech reception capabilities of people who are both deaf and blind and who are extensively trained in the use of Tadoma are comparable to those of normal subjects listening to speech in a background of low-level noise. The exceptional individuals who become proficient in Tadoma provide proof that it is possible to understand speech solely using haptic cues. However, the majority of deaf-blind individuals never reach that level of proficiency, and the intimacy required for communication with the hand resting on the face of the speaker limits its use in social interactions. Nevertheless, Tadoma has been shown to be superior to any artificial tactile display of speech developed to date and has demonstrated the potential of speech communication through the haptic sense.[6]

Another tactile communication method, more commonly used by individuals who are deaf and blind, is the haptic reception of sign language. This method is generally used by deaf people who acquired sign language before becoming blind. To perceive the signs produced by

The speech reception capabilities of people who are both deaf and blind and who are extensively trained in the use of Tadoma are comparable to those of normal subjects listening to speech in a background of low-level noise. The exceptional individuals who become proficient in Tadoma provide proof that it is possible to understand speech solely using haptic cues.

the signing hand(s), the deaf and blind individual places a hand or hands over the signer's hand(s) and passively traces the motion of the signing hand(s). Not surprisingly, the communication rates obtained with haptic reception of sign language are lower than those achieved with visual reception of signs (1.5 signs per second, compared to 2.5 signs per second), and errors are more common with tactile reception.[7] Nevertheless, this haptic method of communication is effective for deaf and blind individuals who are skilled in its use, and the levels of accuracy achieved make it an acceptable means of communication.

Tactile Aids for Vestibular Impairments

The third sensory system for which tactile signals have been used to compensate for sensory loss is the vestibular system, which is involved in maintaining balance and postural control. When the vestibular system malfunctions due to injury or disease, the self-motion cues it provides that help stabilize our vision when moving are reduced, which results in dizziness, blurred vision, and problems in walking and standing. The primary function of balance prostheses based on tactile signals is to reduce body sway and prevent falls. These devices use microelectromechanical systems (MEMS) sensors such as accelerometers and gyroscopes on the head or torso to measure head or body tilt. The sensors are connected to signal processors that translate the outputs into signals, which activate motors in a tactile display worn on the back. The motors are activated when the angular tilt and angular tilt velocity exceeds some threshold value that indicates an increased risk of falling, and the location of the motors activated indicates the direction of body tilt. The objective of the display is to provide tactile cues that the balance-impaired person can use to reduce the body tilt or excessive swaying that can result in falls.[8]

In most of these applications of tactile sensory substitution systems, users require extensive training in order for the devices to be beneficial. The association between the tactile signal and the associated visual or auditory cue is often arbitrary. Research on tactile sensory processing has provided insight into how this translation from one sensory modality to another can be optimized to reduce training time. For some of these devices, the mapping between the tactile signal and its meaning is fairly intuitive—particularly for spatial information, for which people naturally associate a point on the body with the representation of external space. For example, a tactile signal on the right side of the body in the vestibular prosthesis means the user is leaning too far to the right. In other applications, such as using vibrotactile feedback to enhance the learning of speech reading, the mapping of the tactile signal to the associated lip movements being taught is arbitrary.

Tactile Vocabularies

Braille is an example of a tactile communication system in which the pattern of indentations on the fingertip as it is scanned across braille characters is interpreted as letters or other grammatical symbols, thereby providing people who are visually impaired with access to the written word. Over the years there has been interest in seeing if the skin can be used by anyone as a medium of communication similar to vision and audition, and if so, how to develop a tactile vocabulary for this purpose. In the late 1950s Frank Geldard, a psychologist at the University of Virginia, designed a tactile language, called Vibratese, that consisted of 45 basic elements that varied along three first-order dimensions of vibratory stimulation: the amplitude of the signal, its duration, and the location on the body where the vibration was delivered. These dimensions were selected based on earlier research findings that the frequency and amplitude of vibration on the skin are often confounded, which means that as the amplitude of vibration changes at a constant frequency, both the amplitude and the pitch (perceived frequency) of vibration are perceived to change. This meant that either frequency or amplitude could be used to create a tactile communication system, but not both. The other dimension that did not appear to be useful in a tactile coding system was waveform, for which the skin lacked the processing capacity of the ear, where variations in waveform indicate the timbre of a sound. Geldard used three levels of amplitude (soft, medium, and loud), three durations (0.1, 0.3, and 0.5 second) and five distinct locations across the chest where the vibrating motors were attached to create his tactile language. These 45 elements represented all letters, all numerals, and frequently encountered short words (e.g., *in*, *and*, *the*). After 30 hours of learning the Vibratese alphabet and a further 35-hour training period, one person was able to process 38 five-letter words per minute.[9] This compares very favorably with the performance of people trained to receive Morse code tactually, which is on the order of 18 to 24 words per minute. Although Vibratese was never developed further, research on using the skin to communicate continued and experienced a surge of interest in the early 2000s with the advent of small, inexpensive motors that could be incorporated into consumer electronic devices and wearable displays.

More recent work on tactile communication systems has focused on the creation of tactile words or concepts rather than the presentation of individual letters in the word as in Vibratese. These tactile signals are often referred to as *tactons*, by analogy to visual and auditory icons. This approach takes into account the time it takes to present a word tactually, which will always be much slower than what can be processed visually or auditorily. A typical five-letter English word takes about 0.8 seconds to complete in Vibratese, with a maximum delivery rate of 67 words a minute. If concepts or words are presented rather than the individual letters in the word, then the information content of the message is dramatically increased. It is important that these tactile icons are easy to learn and memorize, and if possible have some intuitive meaning; for example, changing the rhythm of a tacton is readily interpreted as indicating urgency or priority.

When a cell phone vibrates in your pocket, the vibration that you feel can be described in terms of its frequency, amplitude, and duration, as shown in figure 13. We can create different patterns of vibration, or tactons, by varying these properties. A low-frequency, less intense vibration signal could indicate an incoming call from a friend, whereas a higher-frequency, more intense signal could mean a message from your boss. It is easier to remember and identify these tactons if they vary along several dimensions, such as intensity, frequency, and duration, and not just a single dimension.

Some aspects of a vibration are more salient to a user than others; for example, the location on the body that the vibration is delivered to is very easily encoded, so if a tactile display is distributed across the body—perhaps in clothing—then spatial location can be used as a building block to create tactons. Other dimensions, such as intensity, are much more difficult for people to identify; it is generally accepted that no more than three levels of

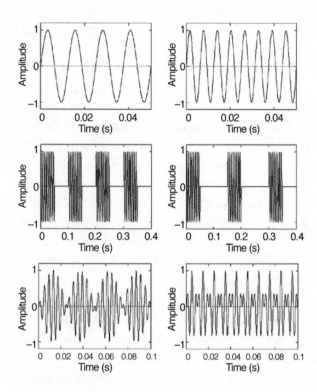

Figure 13 Dimensions of vibrotactile signals that can be used to create tactons. Variations in frequency, the signal duration or repetition rate, and waveform complexity. For the latter a base signal of 250 Hz modulated by a 20 Hz signal (left) feels perceptually rougher than a signal modulated at 50 Hz (right). From Jones and Sarter (2008)[10] with permission of the Human Factors and Ergonomics Society.

intensity should be used in creating tactons. Although simple variations in waveform are difficult to perceive, by modulating the frequency of a base signal it is possible to create tactile patterns that are perceived to vary in roughness. When a base signal of 250 Hz is modulated by a 20 Hz signal it feels perceptually to be rougher than a signal modulated at 50 Hz (see figure 13). This then becomes a dimension that people can effortlessly use to identify a tacton.

Most of the research on tactons has used a vocabulary of around 8 to 15 items, each of which people learn to associate with a particular concept and then are tested on their ability to identify. With this number of tactons and a fairly brief training period, performance is typically around 70% to 80% correct. This suggests that for many applications it will be challenging to create large tactile vocabularies that are easily learned and remembered, unless they are presented in conjunction with visual and/or auditory cues. One reason for this relatively poor performance is that identifying a stimulus in isolation is much harder than comparing two stimuli and determining whether they are the same or different. To put this in context, people can discriminate about 15 steps in intensity between the smallest and the most intense vibration if they are presented in sequence, but only about 3 steps if they have to identify the intensity (i.e., indicate whether it is low, medium, or high). Similarly, the duration of a tacton could span a range of 0.1 to 2 seconds, and across this range about 25 distinct durations can be discriminated but only four or five durations can be identified accurately.

Other aspects of tactile stimulation have been explored in efforts to see how they could be incorporated into communication systems. There has been particular interest in analyzing how the time at which successive tactile stimuli are presented on the skin affects their perceived location and movement across the skin. As described in chapter 4, these effects are illusory, but they provide a mechanism for creating tactile motion on the skin that can be used as a cue in tactons.

SURFACE HAPTICS

With the advent of devices that provide no tactile feedback in situations where we were used to experiencing tactile cues, such as typing on a keyboard, a new term emerged to address this shortcoming. *Surface haptics* refers to the creation of virtual haptic effects on physical surfaces, such as direct-touch user interfaces. This chapter reviews the technologies that have been used to create variable friction displays and describes the challenges associated with modulating the friction experienced on the fingertip as it moves across a flat surface. Incorporating such tactile feedback in flat screen devices is critical to their use as effective human interfaces.

On the now ubiquitous tablets and smartphones that have touch screens, tapping a button or a key on the screen is easily executed without the need for an intervening interface like a physical keyboard or computer mouse. The absence of any haptic feedback on these devices is often problematic because the success of the action cannot be perceived immediately, and because features presented visually cannot be felt. Many of these touch screens also support gestures, making it easy, for example, to navigate through material quickly by swiping the hand across the surface and then focus on particular content by extending the index finger and thumb to zoom into a region of interest.

A new class of displays has emerged to address the limitations of flat-screen devices. They operate by controlling the friction force or lateral resistance to motion between the fingertips and the flat surface of the display. Such devices are often referred to as *variable friction displays*. The interest in incorporating these effects into existing technology is driven in part by the relative ease with which variable friction displays can be integrated into existing tablet and smart-phone screens. The objective of these friction devices is to create the illusion of texture or surface features by varying the lateral forces experienced on the fingertip as it moves across the flat screen. This means that the display can both visually and haptically present features, such as a slider that resists movement and is then released as it is unlocked, or a textured button that can "grasp" a finger to indicate that it has been selected. In contrast to some of the other display technologies described in chapter 5, surface haptic displays are usually co-located with visual displays.

A new class of displays has emerged to address the limitations of flat-screen devices. They operate by controlling the friction force or lateral resistance to motion between the fingertips and the flat surface of the display.

Two main technologies are being explored to create variable friction surface haptics: ultrasonic vibrations and electrovibration, both of which can be used to modulate the perceived friction between the fingertip and the surface of the touch screen. The effects of these two technologies differ in that ultrasonic vibration reduces the perceived friction on a surface, whereas electrovibration enhances it; both require sensing the position of the finger on the display surface. Ultrasonic vibration devices use piezoelectric actuators to vibrate a surface at ultrasonic frequencies (e.g., ~30 kHz) thereby reducing the contact time of the finger on the surface, and in so doing reducing friction. The ultrasonic vibration itself is not perceived, but the decrease in friction between the finger and surface as the amplitude of the vibration increases (~ ±3 μm) is perceptible. Measurements made of the friction force under these conditions show that the friction experienced by the finger moving across the surface can be reduced by up to 95%.

Two mechanisms have been proposed to account for the reduction in friction. First, there is intermittent contact between the finger and the vibrating screen as the finger moves out of phase with the surface of the plate in a bouncing motion; and second, the finger bounces against the squeeze film of pressurized air trapped between the finger and the glass plate. Images taken of the fingertip in contact with the vibrating glass plate indicate that squeeze film levitation is essential to the reduction in friction. This work has also highlighted the importance of the microstructure of the fingertip skin, in particular the asperities, in determining the dynamics of friction—the skin appears to bounce on the surface but does not completely detach from it.[1]

Electrovibration, or electrostatic friction modulation, was first described in the 1950s after the accidental discovery that moving a dry finger over a conductive surface covered with a thin insulating layer excited by a 100 V signal created a rubbery feeling.[2] This sensation results from the frictional shear force created by the electrostatic attraction between the conductive surface and the finger, as illustrated in figure 14. By controlling the amplitude and frequency of the voltage applied to the conductive surface,

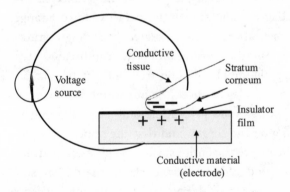

Figure 14 Principle of electrovibration. Redrawn from Giraud et al. (2013)[3] with permission of IEEE.

different textures can be generated. This technology should be distinguished from the electrotactile cutaneous displays described in chapter 5, in which current passing through the skin electrically stimulates the underlying nerve fibers.

Electrovibration has several properties that make it attractive for consumer electronic devices, namely quiet operation, relative ease of implementation with existing touch-screen devices, and the capability of using it on both flat and curved surfaces. Such devices have been developed commercially using self-capacitive touch-screen panels as the display technology. During the past decade, TeslaTouch (Disney Research) and E-Sense Technology (Senseg) used such touch screens, with a transparent electrode on a clear substrate, to electrically induce attractive forces between a moving finger and the surface of the display.[4] Using modulation of the attractive force, a range of tactile sensations were generated, including textures and edges. Specific textural properties can be presented by varying the amplitude and frequency of the signals exciting the display surface. It is important to remember that these effects are perceptible only when the fingers are moving on the screen, and that the same tactile signal is delivered to the entire surface, so different features cannot be produced simultaneously on the display surface. It is possible to design and manufacture systems wherein different areas of the display are independently controlled, which would permit the presentation of more complex tactile patterns. However, these have proven to be very complex systems to control.

All these surface haptics technologies modulate the friction felt by the finger as it moves across a glass surface, although the actual friction experienced by users can vary significantly. The nature of the surface, the moisture content of the skin in contact with the display, and the thickness of the stratum corneum (the outer layer of the skin) can all influence the electrostatic forces on the finger and hence the perceived friction. Moisture content has a substantial effect on friction because it softens the stratum corneum so that it conforms more to the surface, thereby increasing the area of contact. It is possible to couple ultrasonic vibrations with electrovibration to enhance the range of sensations that can be presented to a user; the combination allows both increases and decreases in friction to be displayed on a surface.[5] When this is implemented in a single device, users perceive a friction continuum rather than two distinctly operating effects.

Surface haptics displays have a broad range of applications due to the pervasiveness of flat-screen displays and the need to enhance their tactile communication capabilities. The importance of conveying texture tactually cannot be overestimated in interactive surfaces that are manually explored. For this reason, rendering high-fidelity textures has been the objective of much recent research in this field. A wide array of applications of surface haptics are being explored, from providing navigation cues to the visually impaired, to offering a medium for expressing emotion when communicating virtually,[6] to enhancing educational interactions by engaging the haptic sense in conjunction with visual input.

Surface haptics displays have a broad range of applications due to the pervasiveness of flat-screen displays and the need to enhance their tactile communication capabilities. The importance of conveying texture tactually cannot be overestimated in interactive surfaces that are manually explored.

ARTIFICIAL SENSING: PROSTHETIC AND ROBOTIC HANDS

The principal instrument we use to perceive the world haptically is our hands, and therefore understanding the haptic sense is intimately related to knowing how hands function, as discussed in chapter 2. The sensory and motor properties of the human hand are also relevant to the design of artificial devices such as prosthetic and robotic hands. In this chapter, we examine how prosthetic and robot hands have developed over the years, and the issues associated with imparting a sense of touch to them. Although the technology required for these two applications is fundamentally different, it is interesting to observe that haptic sensing remains a challenging prospect for both.

Because the human hand is the primary structure used to explore the world haptically, it is the standard against which artificial devices are often compared. For prosthetic hands, this makes sense, since the artificial hand is coupled to the body and can make use of the remaining sensory and motor capabilities of the amputated limb and of the higher cortical centers involved in the control of limb movements. There has also been a strong bias in the robotics community toward building anthropomorphic (human-like) robotic hands when dexterous manipulation is required or when it is envisaged that the robotic arm and hand will be controlled by a master device worn by a human operator.[1] Of course, the 200 million years of evolutionary "R and D" that fine-tuned the properties of the human hand should not be overlooked. Nevertheless, as noted by the comparative anatomist Wood Jones, "it is not the hand that is perfect, but the whole nervous mechanism by which movements of the hand are evoked, co-ordinated and controlled."[2] For both prosthetic and robotic hands the benchmarks for performance are often derived from the human hand.

Prosthetic Hands

The development of artificial hands dates back many years. One of the first recorded prosthetic hands was that of the Roman general Marcus Sergius, who was fitted with an iron-made prosthesis after losing his hand during the second Punic war in 218 BCE. This enabled him to return to the battlefield, holding a shield in the prosthesis. In the sixteenth century, the French military surgeon Ambroise Paré designed and built an artificial hand that enabled different types of grasps as well as independent movement of the fingers, using a system of levers and gears. This prosthetic hand was controlled by movement of the opposite limb or the chest. Many years later, in the 1940s, body-powered prostheses were developed in response to the high incidence of war-related upper-extremity amputations. Body-powered prostheses are usually cable-driven via a harness that is strapped around the shoulder of the intact arm and then travels down the prosthetic arm to the end effector, which can be used to grasp objects. The end effector may be a mechanical hook or appendage that is shaped like a hand with one or two movable fingers (see figure 15). The cable control system uses movements and forces generated primarily by shoulder flexion and abduction to control the elbow joint and artificial hand. The orientation of the hand can be controlled by rotating a friction joint at the wrist using the intact hand. Since the majority of upper-extremity amputees have an intact arm, the prosthetic device is not usually used for very precise movements like picking up small objects, but mainly for stabilizing or positioning objects.

Despite their rather primitive structure, body-powered prostheses have remained popular over the years, partly because of their ease of use and the incidental kinesthetic feedback they provide. As the shoulder moves and the cable is pulled through the harness, users come to associate the extent of shoulder movement with the elbow angle in the prosthetic arm. Forces acting on the hand are in turn transmitted through the cable to the body and are perceived remarkably accurately, as reflected in the ability of amputees to duplicate small grasping forces using these devices.[3] The disadvantages of body-powered prostheses are that the configuration of the harness often necessitates extreme body movements, such as rotating the torso in order to turn a heavy object, and that it is difficult to operate the end effector when it is positioned behind the back.

In the late 1940s and early 1950s we saw the development of prosthetic hands controlled by the electrical signals originating in the muscles in the stump of the amputated limb, called myoelectrically controlled hands. Electromyographic (EMG) activity was recorded with electrodes on the skin surface over the muscles within the residual limb, and used to control an electrically powered end effector (see figure 15). These systems varied with respect to the number of electrode sites used for control, the movements that could be executed by the prosthetic hand, and the feedback systems implemented. Although the movements executed by myoelectrically controlled prosthetic hands are considerably slower than those made by the intact hand, the maximum pinch forces they generate can equal, and for some devices exceed, those produced by the intact human hand.

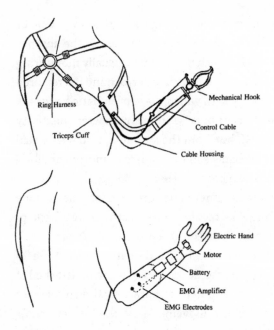

Figure 15 Upper: Bowden cable control system with a mechanical hook. Lower: Myoelectrically controlled below-elbow prosthesis with an electromechanical hand actuated by EMG signals. From Billock (1986)[4] with permission of the American Orthotic and Prosthetic Association.

Myoelectric prostheses are now used by many upper-extremity amputees in the developed world and are particularly useful for those with below-elbow amputations, for whom they have a high level of acceptance. They have a greater range of functional movement than cable-controlled systems and require less effort to operate. However, they are not faster than cable-driven systems, and for many activities of daily living they are slower. One of the major issues associated with their use is the absence of a sense of touch, which necessitates visually monitoring the prosthetic hand to ensure it is performing the required task. The dexterity with which these prosthetic hands are used at present does not arise from any inherent dexterity in the hand itself, but from the amputee, who uses visual cues and subtle changes in the sounds of motors or vibrations to guide the hand.

A considerable amount of research has been done on how to provide tactile and kinesthetic feedback to prosthetic hands, since the absence of such cues places a considerable burden on the user. Research efforts have focused on developing tactile sensors that can be attached to the prosthetic hand to detect contact, and displacement sensors that measure movements of the end effectors. Such sensors have to be durable and robust given the conditions under which they must operate. The signals from these sensors are then processed and presented to the user via electrodes connected to peripheral nerves in the residual limb. The feedback to the amputee may be via an implanted electrode cuff that encircles a nerve so that when electrical pulses are sent by a computer to the cuff, the user feels sensations at particular locations on the hand.[5]

It is clear that as yet there is no prosthesis that can simulate the function and sensibility of the human hand. A number of design constraints related to the acceptable size, weight, and shape of the prosthetic hand have limited the development of a versatile hand. With respect to the weight of a prosthetic arm, it is important to note that although a typical arm, from the midpoint of the humerus bone in the upper arm to the hand, weighs around 3 kg, a prosthetic arm that weighs in excess of 1 or 1.5 kg is considered too heavy to wear by most amputees. This means that less than half of the normal weight is available to a designer to replicate the performance capacities of the human hand. In addition, for users to find it acceptable any control system implemented has to be relatively easy to learn to use, and once mastery is acquired, the prosthetic hand cannot require a high level of concentration to operate.

Recent research on the development of fluidically powered soft actuators opens up exciting possibilities for the future of prosthetic hands. This technology enables deformable digits with embedded tactile sensors to be fabricated for prosthetic hands. These devices have powerful motors and versatile sensing capabilities, as reflected in their capacity to resolve different shapes and surface textures.[6] They are capable of conforming to surfaces or objects and can incorporate rigid elements for structural support or for the integration of components. The use of soft and deformable materials such as these actuators is pushing the boundaries of robotics technologies and their potential areas of application.

It is clear that as yet there is no prosthesis that can simulate the function and sensibility of the human hand. A number of design constraints related to the acceptable size, weight, and shape of the prosthetic hand have limited the development of a versatile hand.

Robotic Hands

The human hand has often been used as the "gold standard" when designing robotic hands because of its dexterity and remarkable sensory capabilities.[1] In many situations, robotic hands use the same tools and manipulate the same objects as human hands and may be controlled remotely by human operators. Anthropomorphic robotic hands therefore make sense in terms of ease of use and controllability. As noted in chapter 1, the human hand is controlled by 38 muscles, has 21 degrees of freedom of movement, and has thousands of sensors embedded in the skin and muscles that provide information about contact, interaction forces, temperature, and vibration. The fingers can work together in close proximity and even overlap each other if needed. Most of the major muscles that control finger movements reside in the forearm, which results in a very compact structure that significantly reduces the mass of the hand itself and provides an impressive workspace for each digit. Replicating these capabilities in a robotic hand is a formidable engineering and scientific challenge, particularly given the constraints on size and mass.

A large variety of robotic hands have been developed over the years, ranging from those that are now used extensively in industry, with claw-like grippers that can grasp objects and perform simple manipulations or highly repetitive actions, to anthropomorphic hands with a number of articulated fingers, capable of grasping and manipulating a diverse range of objects. The latter were initially built as research tools to advance our understanding regarding the control and design of multifingered hands. There are now commercially available robotic hands (e.g., the Barrett hand)[7] that incorporate many of the features of these early research prototypes in a single hand that has the flexibility to operate in a variety of domains. Their use extends from handling hazardous materials, such as nuclear and biohazardous waste, to remote manipulation and parts assembly.

Dexterous robotic hands are typically powered using high-torque, low-mass brushed DC motors, and have three or four articulated fingers, often with one digit placed in opposition to the other fingers (like a thumb) to enable stable grasping. Each finger has three or four degrees of freedom of movement, and in some designs a digit can rotate around a base to enable a greater range of grips. Force/torque and displacement sensors measure the forces produced and movements of the joints. Precision measurement of joint angles allows the accuracy with which the position of robotic fingers can be controlled to exceed that of human fingers. In addition, a few of these hands are capable of moving considerably faster than their human counterparts. Some of the earlier robotic hands, like the Utah/MIT hand designed and built in the 1980s, were developed with the objective of using them in teleoperated environments, and so various types of force-reflecting hand masters (i.e., haptic displays) were built concurrently. Such systems have the advantage that task planning and correcting for errors are the responsibility of the human controlling the hand master. This makes the robotic device computationally less expensive since the controller need not be included in the device.

A more recent example of such a teleoperated system that has achieved commercial success is the da Vinci Surgical System (Intuitive Surgical, Inc.). With this robotic device, a surgeon uses a telemanipulator to perform movements, which are reproduced by robotic arms holding surgical instruments that operate on a patient.[8] In this system, any undesirable features, such as tremor in the surgeon's hands, can be filtered out. The precision with which such movements can be executed and controlled by the robotic arms surpasses that of the surgeon. The da Vinci Surgical System does not at present include any tactile or haptic feedback, although there is ongoing research on how best to incorporate these features in its design.

The recent growth in soft robotics technologies and devices has engendered a new class of robotic hands and grippers that exploit the inherent compliance of the materials to create adaptable systems.[9] Soft grippers can readily deform to the shape of the object being grasped, so they are able to hold a much greater range of objects than rigid grippers can. Similarly, variable-stiffness technologies, in which granular material encased in an elastic membrane can change from very soft to completely stiff when a vacuum is applied, have been used to create grippers that adapt to the shape of objects and then alter their stiffness to hold them securely. Because these soft-body robots are able to stretch, squeeze, and change shape, they open new avenues of application for robotics and human-robot interactions.

Effective grasping by robotic hands requires sensing the contact conditions on the hand, as with human hands, and so tactile sensors that measure pressure and skin deformation are important to their functionality. There have been substantial advances in tactile sensing technology over the past 20 years, and this is reflected in the sensing capabilities of robot hands. However, the development of robotic tactile sensors has not matched the level of sophistication of robotic visual systems, which are used widely in industrial and mobile robot applications. A variety of technologies have been used to create tactile sensors, including capacitive, magnetic, piezoresistive, piezoelectric, and optical, each of which has advantages and limitations. One of the main limitations of many of these sensors is their fragility. When they are attached to fingertips they must be sufficiently robust to withstand the repeated impacts and abrasions associated with making contact with and manipulating objects made from a variety of materials.

The focus in many applications of robotic tactile sensing has been on detecting contact and movement during

The recent growth in soft robotics technologies and devices has engendered a new class of robotic hands and grippers that exploit the inherent compliance of the materials to create adaptable systems. Soft grippers can readily deform to the shape of the object being grasped, so they are able to hold a much greater range of objects than rigid grippers can.

manipulation of an object, and measuring the distribution of normal and tangential forces across the hand.[10] For rigid grippers and robotic hands, tactile array sensors distributed across the fingers and palm enable the contact force and area to be measured and controlled, which is important for detecting slippage of objects and for monitoring interaction forces. In contrast to the speed and accuracy of robotic finger movements, which surpass those of the human hand, the tactile sensing capabilities of all robotic hands are still markedly inferior to those of the human hand. The development of multipurpose sensors that can be embedded in compliant artificial skin surfaces remains a technological challenge to the field of robotic tactile sensing. The soft materials and deformable structures being developed for soft robots require novel approaches to providing sensing capabilities. The emergence of flexible and stretchable electronics and sensors offers exciting opportunities for enhancing tactile sensing in robotic hands.

CONCLUSIONS

Haptic sensing is fundamental to our experience of the physical world, and in its absence we are profoundly limited in our abilities to function—as illustrated vividly by the case of the individual IW, described in chapter 2. Nevertheless we are often unaware of its importance until we are in some way incapacitated by injury or disability. Even then we may comment more on the loss of motor function—that is, that our hands are clumsy—rather than our impaired tactile abilities. Most of the time we are unaware of the continuous tactile stimulation we get through contact with the clothes we wear or the chair supporting us, and it is only when something unexpected occurs that our attention is drawn to our skin. This highlights one of the affordances of tactile and haptic sensing: that we can respond to incoming information rapidly.

Human perception is inherently multisensory; multiple senses are engaged simultaneously as we perceive the world. In our daily lives, we automatically combine information from our senses of touch and kinesthesia with that arising from our other senses. In deciding whether a strawberry we see (visual sensing) at the market is ripe, for example, we may initially squeeze it (haptic sensing), then smell it (olfactory sensing), and finally bite into it (gustatory sensing) to confirm that our sense of taste is consistent with our visual, haptic, and olfactory impressions. The information we get from our various senses may be redundant or complementary. There are situations where we defer to the haptic sense as the superior sense in making a judgment, such as evaluating the feel of a fabric, the finish on a piece of furniture, or the weight of a fruit. In other circumstances, we may simply want to confirm haptically what we have seen or heard. The bidirectional nature of haptic sensing, in which we not only perceive the world but act directly on it to obtain those sensory impressions, makes it unique among our senses. When we look at how people use their hands to discover the properties of objects they encounter, we find remarkable consistency in the types of movements used. Moreover, when a specific property is of interest, for example whether a piece of fruit is ripe, the movement is inevitably optimal for perceiving that property (i.e., compliance).

In deciding whether a strawberry we see (visual sensing) at the market is ripe, for example, we may initially squeeze it (haptic sensing), then smell it (olfactory sensing), and finally bite into it (gustatory sensing) to confirm that our sense of taste is consistent with our visual, haptic, and olfactory impressions.

Tactile display technology has a long history, dating back to the development of braille and other sensory substitution systems that were designed to compensate for sensory loss. More recently, vibrotactile displays have become a ubiquitous feature in smartphones, tablets, and other small electronic devices. The fact that such communication is private and unobtrusive has obvious appeal. For most of these devices, tactile communication has been limited to providing the user with alerts and notifications. Although some of these systems allow a much greater range of tactile communication capabilities by customizing the inputs delivered, such functionality has not been widely adopted. These features may be considered less useful because of the limits of our haptic information processing capacity, compared to vision and audition, and the requirement that users focus their attention on the tactile signal to determine its meaning. With the advent of surface haptic displays based on variable friction technologies, we see this limitation overcome, in that such displays involve both the visual and haptic senses. As surface haptic display technologies mature, high-fidelity textures will be rendered, markedly enhancing our interactions with a broad range of devices.

The use of the sense of touch as a navigation aid seems to be one of the most promising general applications of tactile displays. The reports of people walking into lampposts or into traffic while looking at a cell phone for directions would become less frequent if the directions were provided by smart shoes or clothing equipped with vibrating motors and sensors that communicated with a phone's navigation app. The use of tactile feedback to convey information about posture seems particularly promising for athletes and fitness trainers. In these applications, directional tactile feedback can be used in conjunction with an accompanying app to indicate which joint to move or how to correctly align the position of the body for a specific activity. In this sphere of wearables and interactive clothing, there is potential for considerable growth.

Haptic interfaces in which the user makes contact with an object in a real or virtual environment, and feels its properties such as weight or stiffness, have been evaluated in a diverse range of applications. They have been used in areas such as medical and dental training scenarios, prototyping, gaming, education, and assistive technology, but as yet have not been widely adopted. This is partly due to the cost of such systems, the perceived value associated with incorporating haptic interfaces in larger systems, and the absence of any multipurpose display that can readily be adapted for different applications. Future haptic displays will require highly realistic haptic rendering so that the user experiences sensations similar to those afforded by direct manual interactions with the environment.

In an era when so much communication takes place via the Internet and without any direct person-to-person contact, the significance of the loss of haptic communication, whether a handshake, a pat on the shoulder, or a gentle caress, cannot be overestimated. Interpersonal touch has been shown to affect people's attitudes toward others who are delivering services and can facilitate bonding between people. To address this shortcoming of modern communication systems, a number of simple devices have been developed over the past decade to enable long-distance tactile communication. These include a ring that squeezes the finger and a jacket that "hugs" the wearer when activated. Such devices have yet to meet commercial success, and so it is not clear whether the absence of any rigorous empirical testing prior to device development has impeded their adoption. In developing such devices it will be critical to identify the type of tactile experience that people desire for long-distance interpersonal communication, making sure to rely as much on our knowledge of human haptic perception as on the availability of new technologies.

Many fundamental questions remain to be answered about haptic perception, about the underlying sensory mechanisms involved in processing information from the skin and the muscles, and about how different sensory modalities interact. Still, we have seen remarkable

In an era when so much communication takes place via the Internet and without any direct person-to-person contact, the significance of the loss of haptic communication, whether a handshake, a pat on the shoulder, or a gentle caress, cannot be overestimated.

advances over the past 20 years in our understanding of the human haptic system and in the development of technologies enabling us to study it more effectively. The importance of haptic perception to our daily interactions must not be underestimated, and its absence in many spheres of activity presents a challenge that will have to be surmounted.

NOTES

Chapter 1

1. McGlone, F., & Reilly, D. (2010). The cutaneous sensory system. *Neuroscience and Biobehavioral Reviews*, *34*, 148–159.

2. Linden, D. J. (2015). *Touch: The Science of Hand, Heart, and Mind*. New York: Viking Press.

3. Napier, J. R. (1993). *Hands* (revised by R. H. Tuttle). Princeton, NJ: Princeton University Press.

4. Johnson, K. O. (2001). The roles and functions of cutaneous mechanoreceptors. *Current Opinions in Neurobiology*, *11*, 455–461.

5. Gescheider, G. A., Wright, J. H., & Verrillo, R. T. (2009). *Information-Processing Channels in the Tactile Sensory System*. New York: Taylor & Francis.

6. McGlone, F., Olausson, H., Boyle, J. A., Jones-Gotman, M., Dancer, C., Guest, S., & Essick, G. (2012). Touching and feeling: Differences in pleasant touch processing between glabrous and hairy skin in humans. *European Journal of Neuroscience*, *35*, 1782–1788.

7. Jones, L. A., & Smith, A. M. (2014). Tactile sensory system: Encoding from the periphery to the cortex. *WIREs System Biology and Medicine*, *6*, 279–287.

Chapter 2

1. Proske, U., & Gandevia, S. C. (2012). The proprioceptive senses: Their roles in signaling body shape, body position and movement, and muscle force. *Physiological Reviews*, *92*, 1651–1697.

2. Cole, J. (1995). *Pride and a Daily Marathon*. Cambridge, MA: MIT Press.

3. Cole, J. (2016). *Losing Touch: A Man Without His Body*. New York, NY: Oxford University Press.

4. Jones, L. A. (2003). Perceptual constancy and the perceived magnitude of muscle forces. *Experimental Brain Research*, *151*, 197–203.

5. Buchthal, F., & Schmalbruch, H. (1980). Motor unit of mammalian muscle. *Physiological Reviews*, *60*, 90–142.

6. Diamond, M. E., von Heimendahl, M., & Arabzadeh, E. (2008). Whisker-mediated texture discrimination. *PLOS Biology*, *6*, 1627–1630.

7. Gerhold, K. A., Pellegrino, M., Tsunozaki, M., Morita, T., Leitch, D. B., Tsuruda, P. R., Brem, R. B., Catania, K. C., & Bautista, D. M. (2013). The

star-nosed mole reveals clues to the molecular nature of mammalian touch. *PLOS One, 8*, e55001.

8. Catania, K. C. (2011). The sense of touch in the star-nosed mole: From mechanoreceptors to the brain. *Philosophical Transactions of the Royal Society of London B; Biological Sciences, 366*, 3016–3025.

9. Stevens, J. C. (1991). Thermal sensibility. In M. Heller & W. Schiff (Eds.), *The Psychology of Touch* (pp. 61–90). Hillsdale, NJ: Erlbaum.

10. Vay, L., Gu, G., & McNaughton, P. A. (2012). The thermos-TRP ion channel family: Properties and therapeutic implications. *British Journal of Pharmacology, 165*, 787–801.

11. Filingeri, D., & Havenith, G. (2015). Human skin wetness perception: Psychophysical and neurophysiological bases. *Temperature, 2*, 86–104.

12. Napier, J. R. (1993). *Hands* (revised by R. H. Tuttle). Princeton, NJ: Princeton University Press.

13. Foucher, G., & Chabaud, M. (1998). The bipolar lengthening technique: A modified partial toe transfer for thumb reconstruction. *Plastic and Reconstructive Surgery, 102*, 1981–1987.

14. Jones, L. A., & Lederman, S. J. (2006). *Human Hand Function*. New York: Oxford University Press.

Chapter 3

1. Jones, L. A., & Lederman, S. J. (2006). *Human Hand Function*. New York: Oxford University Press.

2. Verrillo, R. T., Bolanowski, S. J., & McGlone, F. P. (1999). Subjective magnitude estimate of tactile roughness. *Somatosensory and Motor Research, 16*, 352–360.

3. Brodie, E. E., & Ross, H. E. (1985). Jiggling a lifted weight does aid discrimination. *American Journal of Psychology, 98*, 469–471.

4. Skedung, L., Arvidsson, M., Chung, J. Y., Stafford, C. M., Berglund, B., & Rutland, M. W. (2013). Feeling small: Exploring the tactile perception limits. *Scientific Reports, 3*, 2617.

5. Gallace, A., & Spence, C. (2014). *In Touch with the Future*. New York: Oxford University Press.

6. Peters, R. M., Hackeman, E., & Goldreich, D. (2009). Diminutive digits discern delicate details: Fingertip size and the sex difference in tactile spatial acuity. *Journal of Neuroscience, 29*, 15756–15761.

7. Gescheider, G. A., Bolanowski, S. J., Pope, J., & Verrillo, R. T. (2002). A four-channel analysis of the tactile sensitivity of the fingertip: Frequency

selectivity, spatial summation, and temporal summation. *Somatosensory & Motor Research*, *19*, 114–124.

8. Hollins, M., Faldowski, R., Rao, S., & Young, F. (1993). Perceptual dimensions of tactile surface texture: A multidimensional scaling analysis. *Perception & Psychophysics*, *54*, 697–705.

9. Jones, L., Hunter, I., & Lafontaine, S. (1997). Viscosity discrimination: A comparison of an adaptive two-alternative forced-choice and an adjustment procedure. *Perception*, *26*, 1571–1578.

10. Lederman S. J., & Klatzky, R. L. (1987). Hand movements: A window into haptic object recognition. *Cognitive Psychology*, *19*, 342–368.

11. Van Polanen, V., Bergmann Tiest, W. M., & Kappers, A. M. L. (2012). Haptic pop-out of moveable stimuli. *Attention, Perception, & Psychophysics*, *74*, 204–215.

Chapter 4

1. Lederman, S. J., & Jones, L. A. (2011). Tactile and haptic illusions. *IEEE Transactions on Haptics*, *4*, 273–294.

2. Heller, M. A., Brackett, D. D., Wilson, K., Yoneama, K., Boyer, A. I., & Steffen, H. (2002). The haptic Müller-Lyer illusion in sighted and blind people. *Perception*, *31*, 1263–1274.

3. Kappers, A. M. (1999). Large systematic deviations in the haptic perception of parallelity. *Perception*, *28*, 1001–1012.

4. Goldreich, D. (2007). A Bayesian perceptual model replicates the cutaneous rabbit and other spatiotemporal illusions. *PLOS One*, *2*, e333.

5. Geldard, F. A., & Sherrick, C. E. (1972). The cutaneous "rabbit": A perceptual illusion. *Science*, *178*, 178–179.

6. Botvinick, M., & Cohen, J. (1998). Rubber hands "feel" touch that eyes see. *Nature*, *391*, 756.

7. Craske, B. (1977). Perception of impossible limb positions induced by tendon vibration. *Science*, *196*, 71–73.

8. Gandevia, S. C., & Phegan, C. M. L. (1999). Perceptual distortions of the human body image produced by local anesthesia, pain and cutaneous stimulation. *Journal of Physiology*, *514*, 609–616.

9. Robles-De-La-Torre, G., & Hayward, V. (2001). Force can overcome object geometry in the perception of shape through active touch. *Nature*, *412*, 445–448.

10. Srinivasan, M. A., Beauregard, G. L., & Brock, D. L. (1996). The impact of visual information on the haptic perception of stiffness in virtual

environments. *Proceedings of the ASME Dynamic Systems and Control Division*, *58*, 555–559.

11. Lécuyer, A. (2009). Simulating haptic feedback using vision: A survey of research and applications of pseudo-haptic feedback. *Presence*, *18*, 39–53.

Chapter 5

1. Jones, L. A., & Sarter, N. B. (2008). Tactile displays: Guidance for their design and application. *Human Factors*, *50*, 90–111.

2. Bach-y-Rita, P., Kaczmarek, K., Tyler, M., & Garcia-Lara, J. (1998). Form perception with a 49-point electrotactile stimulus array on the tongue. *Journal of Rehabilitation Research and Development*, *35*, 427–430.

3. Inoue, S., Makino, Y., & Shinoda, H. (2015). Active touch perception produced by airborne ultrasonic haptic hologram. *IEEE World Haptics Conference*, 362–367.

4. Gleeson, B. T., Horschel, S. K., & Provancher, W. R. (2010). Design of a fingertip-mounted tactile display with tangential skin displacement feedback. *IEEE Transactions on Haptics*, *3*, 297–301.

5. Ho, C., Reed, N., & Spence, C. (2007). Multisensory in-car warning signals for collision avoidance. *Human Factors*, *49*, 1107–1114.

6. Rupert, A. H. (2000). An instrumentation solution for reducing spatial disorientation mishaps: A more "natural" approach to maintaining spatial orientation. *IEEE Engineering in Medicine and Biology Magazine*, March/April, 71–80.

7. Burdea, G. C. (1996). *Force and Touch Feedback for Virtual Reality*. New York: Wiley.

8. Brooks, F., Ouh-Young, M., Batter, J., & Jerome, A. (1990). Project GROPE: Haptic displays for scientific visualization. *Computer Graphics*, *24*, 177–185.

9. Immersion Corporation: https://immersion.com.

10. Sensable Technologies, acquired by Geomagic in 2012, acquired by 3D Systems in 2013: https://www.3dsystems.com.

11. Haption: https://haption.com.

12. Force Dimension: http://forcedimension.com.

13. Butterfly Haptics: http://butterflyhaptics.com.

14. Selzer, R. (1982). *Letters to a Young Doctor*. New York: Simon & Schuster.

Chapter 6

1. Reed, C. M., & Durlach, N. I. (1998). Note on information transfer rates in human communication. *Presence*, *7*, 509–518.

2. Bliss, J. C., Katcher, M. H., Rogers, C. H., & Shepard, R. P. (1970). Optical-to-tactile image conversion for the blind. *IEEE Transactions on Man-Machine Systems*, *11*, 58–65.

3. White, B. W., Saunders, F. A., Scadden, L., Bach-Y-Rita, P., & Collins, C. C. (1970). Seeing with the skin. *Perception & Psychophysics*, *7*, 23–27.

4. Nau, A., Bach, M., & Fisher, C. (2013). Clinical tests of ultra-low vision used to evaluate rudimentary visual perceptions enabled by the BrainPort vision device. *Translational Vision Science & Technology*, *2*, doi:10.1167/tvst.2.3.1.

5. Reed, C. M., & Delhorne, L. A. (1995). Current results of a field study of adult users of tactile aids. *Seminars in Hearing*, *16*, 305–315.

6. Reed, C. M. (1995). Tadoma: An overview of research. In G. Plant & K.-E. Spens (Eds.), *Profound Deafness and Speech Communication* (pp. 40–55). London: Whurr Publishers.

7. Reed, C. M., Delhorne, L. A., Durlach, N. I., & Fischer, S. D. (1995). A study of the tactual reception of sign language. *Journal of Speech and Hearing Research*, *38*, 477–489.

8. Wall, C. III, Weinberg, M. S., Schmidt, P. B., & Krebs, D. E. (2001). Balance prosthesis based on micromechanical sensors using vibrotactile feedback of tilt. *IEEE Transactions on Biomedical Engineering*, *48*, 1153–1161.

9. Geldard, F. A. (1957). Adventures in tactile literacy. *American Psychologist*, *12*, 115–124.

10. Jones, L. A., & Sarter, N. B. (2008). Tactile displays: Guidance for their design and application. *Human Factors*, *50*, 90–111.

Chapter 7

1. Wiertlewski, M., Friesen, R. F., and Colgate, J. E. (2016). Partial squeeze film levitation modulates fingertip friction. *Proceedings of the National Academy of Sciences*, *113*, 9210–9215.

2. Mallinckrodt, E., Hughes, A., & Sleator, W. (1953). Perception by the skin of electrically induced vibrations. *Science*, *118*, 277–278.

3. Giraud, F., Amberg, M. & Lemaire-Semail, B. (2013). Merging two tactile stimulation principles: Electrovibration and squeeze film effect. *Proceedings of the World Haptics Conference*, 199–203.

4. Bau, O., Poupyrev, I., Israr, A., & Harrison, C. (2010). TeslaTouch: Electrovibration for touch surfaces. *Proceedings of the 23rd Annual ACM Symposium on User Interface Software and Technology*, 283–292.

5. Vezzoli, E., Messaoud, W. B., Amberg, M., Giraud, F., Lemaire-Semail, B., & Bueno, M.-A. (2015). Physical and perceptual independence of ultrasonic

vibration and electrovibration for friction modulation. *IEEE Transactions on Haptics*, 8, 235–239.

6. Mullenbach, J., Shultz, C., Colgate, J. E., & Piper, A. M. (2014). Exploring affective communication through variable-friction surface haptics. *Proceedings of the ACM SIGCHI Conference on Human Factors in Computing Systems*, 3963–3972.

Chapter 8

1. Balasubramanian, R., & Santos, V. J. (Eds). (2014). *The Human Hand as an Inspiration for Robot Hand Development.* New York: Springer.

2. Wood Jones, F. (1944). *The Principles of Anatomy as Seen in the Hand* (2nd ed.). London: Bailliere, Tindall.

3. Jones, L. A. (1997). Dextrous hands: Human, prosthetic, and robotic. *Presence*, 6, 29–56.

4. Billock, J. N. (1986). Upper limb prosthetic terminal devices: Hands versus hooks. *Clinical Prosthetics and Orthotics*, 10, 57–65.

5. Tyler, D. J. (2016). Restoring the human touch. *IEEE Spectrum*, May, 28–33.

6. Zhao, H., O'Brien, K., Li, S., & Shepherd, R. F. (2016). Optoelectronically innervated soft prosthetic hand via stretchable optical waveguides. *Science Robotics*, 1, eaai7529.

7. Barrett Technology, Inc.: http://www.barrett.com.

8. Da Vinci Surgical System: http://www.davincisurgery.com/da-vinci-surgery/da-vinci-surgical-system.

9. Laschi, C., Mazzolai, B., & Cianchetti, M. (2016). Soft robotics: Technologies and systems pushing the boundaries of robot abilities. *Science Robotics*, 1, eaah3690.

10. Cutkosky, M., Howe, R. D., & Provancher, W. R. (2008). Force and tactile sensors. In B. Siciliano & O. Khatib (Eds.), *Springer Handbook of Robotics* (pp. 455–476). Berlin: Springer.

GLOSSARY

afferent
A nerve fiber going from the periphery, such as the skin, toward the spinal cord and the brain.

carpals
Eight bones arranged in two rows that form the wrist; the proximal row articulates with the radius and ulna bones in the forearm, and the distal row articulates with the metacarpal bones in the palm of the hand.

corollary discharges
Copies or correlates of the descending motor signals sent from the brain to the spinal cord; they are transmitted to the sensory areas in the brain.

difference threshold
Smallest amount of change in a stimulus required to produce a perceptible change in sensation.

dynamic range
Ratio of the largest to the smallest intensity of a stimulus, measured in decibels (dB).

eccrine sweat glands
Glands of the body that are involved in temperature regulation through secretion of sweat.

efferent
Outgoing signals, for example, from the brain to the muscles.

epidermis
Outermost layer of skin; provides a protective barrier, preventing loss of moisture and entry of bacteria.

glabrous skin
Hairless skin found on the palms of the hands and soles of the feet that is thicker than hairy skin; in the hand it bends along the flexure lines when an object is grasped.

Golgi tendon organ
An encapsulated receptor that senses force, normally found at the junction between muscle tendons and muscle fibers.

guard hairs
Longer, thicker hairs found in hairy skin. More visible than the fine vellus hairs.

JND
Just noticeable difference, the physical difference between two stimuli that is required to achieve a just noticeable difference in sensation.

kinesthesia
Sense of limb position and movement.

mechanoreceptors
Receptors in skin that respond to mechanical pressure or stretch.

Meissner's corpuscles
Encapsulated tactile receptors that are rapidly adapting and found in the superficial layers of glabrous skin on the palms of the hand and soles of the feet; they are sensitive to changes in pressure.

Merkel cells
Tactile receptors located in clusters at the tip of the epidermal folds that project into the dermis.

metacarpals
The five bones in the palm of the hand.

nociception
Sense of pain, mediated by pain receptors (nociceptors) that respond to several different forms of stimulation—mechanical, electrical, chemical, and thermal—that can damage tissue.

Pacinian corpuscles
Ovoid tactile receptors located deep within the dermis and subcutaneous fat layer.

papillary ridges
Thickening of the epidermis, raised above the skin and found on the palmar surface of the hands and the soles of the feet.

phalanges
The fourteen bones that make up the digits; there are two in the thumb and three in each finger.

Phi phenomenon
Illusion that occurs when a number of discrete taps are delivered to the skin and are experienced as a single stimulus moving across the skin rather than as isolated taps.

proprioception
Sense of limb position and movement, mediated by receptors in muscles, skin, and joints; sometimes used synonymously with *kinesthesia*.

receptive field
Area of a receptive sheet, such as the skin or the retina in the eye, within which activity of a neuron can be influenced.

Ruffini endings
Spindle-shaped tactile receptors, located in the connective tissue of the dermis.

saddle joint
Carpometacarpal joint at the base of the thumb; it permits flexion/extension and abduction/adduction movements in addition to the axial rotation of the thumb that brings it into contact with the fingertips.

sensory saltation
Spatial tactile illusion in which a series of short pulses delivered at three different locations on the skin are perceived as a stimulus that is moving progressively across the skin.

slowly adapting unit
A tactile afferent unit that has a sustained response to indentation of the skin.

stratum corneum
The outermost layer of the epidermis, which varies in thickness from 0.2 to 2 mm; it contains flattened dry cells of soft keratin that are bound together and sloughed off.

surface haptics
Creation of virtual haptic effects such as textures on flat physical surfaces such as direct-touch user interfaces.

tactons
"Words" in a tactile vocabulary, created by varying the properties of a vibrotactile signal, that have an abstract meaning.

Tau effect
The illusory experience that when two stimuli are presented on the skin in quick succession they are perceived to be closer together spatially than if there is a longer interval between them.

thenar eminence
The fleshy region on the palm of the hand at the base of the thumb that overlies the thumb metacarpal.

two-point threshold
Distance at which two points of stimulation on the skin are perceived to be distinct and not a single point.

vellus hairs
Soft, fine hairs in hairy skin that serve a wicking function, drawing perspiration away from the skin, and contribute to thermal insulation.

vestibular system
Sensory system involved in maintaining balance and spatial orientation based on inputs from structures in the inner ear.

Weber's illusion
Phenomenon that a given distance between two points of stimulation on the skin is perceived to be greater on areas of the body with higher tactile spatial acuity, such as the hand.

Weber's law
A principle discovered by Weber that the change in stimulus intensity that can be discriminated is a constant fraction of the level of stimulus intensity. The law has been found to apply across a broad range of sensory modalities. The value is referred to as the *Weber fraction*.

FURTHER READING

Cole, J. (2016). *Losing Touch: A Man Without his Body*. New York, NY: Oxford University Press.

Gallace, A., & Spence, C. (2014). *In Touch with the Future*. New York, NY: Oxford University Press.

Gescheider, G. A., Wright, J. H., & Verrillo, R. T. (2009). *Information-Processing Channels in the Tactile Sensory System*. New York, NY: Taylor & Francis.

Grunwald, M. (2008). *Human Haptic Perception: Basics and Applications*. Basel, Switzerland: Birkhauser.

Jones, L. A., & Lederman, S. J. (2006). *Human Hand Function*. New York, NY: Oxford University Press.

Linden, D. J. (2015). *Touch: The Science of Hand, Heart, and Mind*. New York, NY: Viking Press.

Napier, J. R. (1993). *Hands. Revised by R. H. Tuttle*. Princeton, NJ: Princeton University Press.

Prescott, T. J., Ahissar, E., & Izhikevich, E. (2016). *Scholarpedia of Touch*. Amsterdam, the Netherlands: Atlantis Press.

INDEX